History of Information Security

Series Editor
Karl de Leeuw (Amsterdam)

Advisory Board
John F. Dooley (Knox College)
Joseph Fitsanakis (Coastal Carolina University)
Ioanna Iordanou (Oxford Brookes University)
Benedek Lang (Budapest University of Technology)
Jean-Jacques Quisquater (Université catholique de Louvain)
Betsy Rohaly Smoot (Independent Scholar / Retired NSA Historian)
John Tucker (Swansea University)
Joachim von zur Gathen (Bonn-Aachen International Center
    for Information Technology and Universität Bonn)

The Springer book series History of Information Security publishes monographs about all aspects of the history of cryptology, signals intelligence, computer security, and information assurance, focusing on the interplay between technological change and organizational requirements. There are no limits in time period – monographs dealing with information assurance and warfare in antiquity are as welcome as those dealing with cybercrime – but subjects should be clearly defined, limited, and treated comprehensively. Prospective authors are invited to submit proposals or manuscripts, aimed at academic and/or non-expert audiences.

More information about this series at http://www.springer.com/series/16513

Joseph Fitsanakis

Redesigning Wiretapping

The Digitization of Communications Interception

 Springer

Joseph Fitsanakis
Intelligence and National Security Studies Program
Coastal Carolina University
Conway, SC, USA

ISSN 2662-7558   ISSN 2662-7566   (electronic)
History of Information Security
ISBN 978-3-030-39921-4   ISBN 978-3-030-39919-1   (eBook)
https://doi.org/10.1007/978-3-030-39919-1

This Springer imprint is published by the registered company Springer Nature Switzerland AG.
The registered company address is: Gewerbestrasse 11, 6330 Cham, Switzerland

Preface

The concept of *informatization* was introduced in political research in 1977. In the ensuing 43 years, scholars have become increasingly cognizant of the powerful interaction between information technology and social development. This research follows in the footsteps of this increased awareness and explores the sociotechnical impact of digitization on information security and—by extension—privacy.

This book seeks to contribute to a wider body of literature in information security that desires to provide meaningful answers to the following questions: (1) What sociotechnical trends have been evident in the recent history of information-security policy in the United Kingdom and the United States? (2) What particular political visions do these trends seem to favor and what do they appear to suggest for the future of information security in the West? (3) What is the importance of digital information security for practices of policing, both by governmental decision centers and commercial bodies?

To answer these questions, this book examines the historical emergence of two legislative frameworks: the Communications Assistance for Law Enforcement Act (CALEA), enacted by the United States Congress in 1994, and the Regulation of Investigatory Powers Act (RIPA), enacted by the British Parliament in 2000. The two Acts, which remain active to this day, make it mandatory for telecommunications service providers to, among other things, ensure that their customers' communications can be intercepted by law enforcement and security agencies, whose interception capabilities were seriously hampered by the digitization and deregulation of telecommunications after 1984.

The project rests on previously unpublished qualitative data on RIPA and CALEA. These have been acquired through open-source, restricted, or leaked government and industry reports on the subject, as well as through over 40 in-depth interviews with some of the protagonists representing different sides in the negotiations over RIPA and CALEA, nearly 20 years ago. The development of communications interception as a component of information security is thus examined historically within the context of complex relationships between political actors, such as national policy experts and government advisors, state and corporate decision-makers, and members of regulatory bodies.

Conway, SC   Joseph Fitsanakis
December 2019

Contents

Chapter 1
Introduction

In 1998 I was living in a southern region of the United States. The Internet had just begun to emerge as a civilian utility at that time, with only about 20% of American households accessing the World Wide Web through dial-up services. Few people outside of government or academia used email, while cellular-based text messaging was almost non-existent. Cellular telephones—the unwieldly ancestors of today's mobile phones—were around, but they were few and expensive to use, especially while roaming. Landline telephony was therefore still the dominant form of tele-communication. A small number of politicians had just begun to sense the dawn of the upcoming telecommunications revolution and were contemplating the future of the so-called *information superhighway*. Known also as the *inforbahn*, the term signified a proposed high-speed telecommunications system that would connect the nation with fiber optic networks and lead it to the information society of the twenty-first century.

Late one February night, someone—seemingly a random stranger—called my landline telephone and left a message on my answering machine, in which he threatened to come to my residence and hurt me. The caller did not seem to know who he was contacting. He was almost certainly a bored teenager using his parents' telephone line to prank a random subscriber. It was something like the pre-Internet version of cyber-bullying. I was not especially concerned, but I was not about to take any chances, so I contacted the police to report the incident. When a police officer came to my house, I took her to the answering machine—a bulky stand-alone accessory that users attached to the telephone appliance in order to record messages from callers—and replayed her the threatening message. I imagined that the officer would resolve my case that same evening. All she had to do was complete a one-page report and file it at her precinct station. A junior detective would then contact the designated law enforcement point-of-contact of the local telephone exchange carrier and request a transcript of the call data for my subscriber line. Less than half an hour later, the telephone company would forward the detective a list of telephone numbers that contacted my number during a prescribed period. It would then be an easy task to match the timestamp of the incriminating message left

© Springer Nature Switzerland AG 2020
J. Fitsanakis, *Redesigning Wiretapping*, History of Information Security,
https://doi.org/10.1007/978-3-030-39919-1_1

on my answering machine with the call data supplied by the local telephone exchange carrier, and isolate the incriminating number. Then, again with the help of the local telephone exchange, the detective would identify the name and address of the incriminating subscriber before sending around a police officer to scold the bored teenager. That—I was certain—would be the end of the matter.

I was wrong. Five days later, a detective from the police station called to tell me that the local telephone exchange carrier was not cooperating and that it would be close to impossible to resolve the case without a warrant. Such a request would take time, and might never be granted, said the detective. Even if it were granted, he added, the incriminating telephone number might end up belonging to a public telephone, which would mean a possible dead end in the investigation. But the main obstacle, said the detective, was "the guys at the telephone company". He went on to explain that "until a few months ago, it was an AT&T [American Telephone and Telegraph Company] outfit. We knew them and they knew us. These kinds of subscriber line requests used to take five minutes to take care of. We could even do a quick [wire]tap and see if the prank caller bothered other subscribers, then pursue it as a large-scale public nuisance case", he said. "But now it's just impossible. They use fiber switches and the local carrier has been bought by a company in [New] Jersey. I don't even know how to contact them".

Unbeknownst to the detective, I was acutely aware of what he was referring to. I was an academic researcher studying the use of telecommunications interception by law enforcement, security and intelligence services, and was regularly travelling back and forth across the Atlantic to compare state-authorized telecommunications-interception practices in Britain and the United States. In fact, my unproductive conversation with the frustrated detective lies at the heart of this study, which is about a historic paradigm shift in telecommunications information security.

1.1 Information Security and Communications Interception

Since their early emergence in the 1830s, the success of telecommunications systems has depended on their ability to provide two things, namely efficient communications exchanges between users and secure networks. The latter principle, known as information security, ensures that confidential or otherwise private information is exchanged securely between users. But what happens when users utilize the telecommunications network for illicit purposes? Governments generally view the illicit use of telecommunications systems as a potential threat to their integrity and ability to enforce the law. Historically, such concerns have been addressed through various forms of policing of telecommunications, including state-sponsored telecommunications surveillance. In other words, governments have always demanded access to the telecommunications activity of subscribers, in order to help combat crime and safeguard national security. This access has been granted to them without much resistance by the telecommunications service providers. After all, governmental demands for access to the telecommunications networks were based on state prerogatives that go as far back as the emergence of the first courier-based mail services in the sixteenth century.

By the 1930s, telegraphy had given its way to another form of electric communication, namely telephony. By that time, most countries—whether capitalist or communist—had nationalized their nationwide telephone system. The latter was seen, like the military and the police, as too sensitive for national security to be left to the private sector. A rare exception to that norm was the United States, where telephony did not evolve through nationalization, but rather through fierce competition in the private sector. But even in that case, the American government exercised strict regulatory pressure on the private telecommunications sector and went out of its way to ensure that it had unrestricted access to the network, so as to police it. Through these arrangements, the telecommunications service industry developed close and long-lasting relations with law enforcement and intelligence agencies. As this book will demonstrate, their relationship evolved into an intimate set of both formal and informal agreements that was useful for both parties.

But in the 1980s and early 1990s, this centuries-long state of affairs was abruptly challenged by two largely unforeseen factors. One of them was technological and can be summarized as the onset of *digitization*, namely the process of converting analog information into binary digits, thus making it computer-readable. This new technological paradigm revolutionized the telecommunications landscape by allowing providers to offer new services through advanced, fiber optic networks. The digital delivery platforms led to the emergence of a host of new and exciting telephone features such as speed-dialing, call-forwarding, remote-answering services and voicemail, teleconferencing, and many others, which users grew to rely on. However, these added digital services posed increasingly insurmountable technical barriers to the interception capabilities of governments. The latter had been designed for analog systems that were now quickly evolving into digital networks and—as this book will explain—were immune to interception by government agencies.

The other unforeseen factor that threatened the state's ability to intercept was political. It had to do with the decision by governments to reinvigorate the telecommunications sector by deregulating and privatizing it. In 1979 and 1980, two uncompromising neoliberal 'rock stars', Margaret Thatcher and Ronald Reagan, took Britain and the United States by storm. Their economic policies centered on shrinking the size of government and deregulating the most dynamic sectors of industry, including telecommunications. They moved fast. By 1984 the administration of President Reagan had terminated the monopoly of AT&T, a legendary American institution, which had lasted for nearly 100 years. That same year in Britain, the government of Prime Minister Thatcher began privatizing the most profitable parts of the telecommunications network of British Telecom. Known as BT, the corporation had previously been part of the state-owned General Post Office, which had operated as the British government's communication arm since 1660. There is no question that these seismic changes gave an immediate boost to the economic sectors of both Britain and the United States. These were felt globally through the stock market.

At the same time, however, these revolutionary changes posed seemingly insurmountable challenges to the ability of government agencies to police the telecommunications networks. Not only were these networks turning digital—thus inaccessible by the analog interception systems employed by the agencies—they were also fragmenting into literally thousands of different private companies. In other words, the once intimate institutional interface between the security agencies and the telecommunications sector was becoming vastly more complex. In the words of the frustrated detective who spoke with me in February of 1998, prior to deregulation the local exchange carrier "was an AT&T outfit. We knew them and they knew us. These kinds of subscriber line requests used to take five minutes to take care of". But police and security officers were now finding that different parts of the telephone network were owned by a milieu of different companies. These companies were geographically remote and had differing sizes, cultures, and levels of experience. Most importantly, they had differing degrees of willingness to assist law enforcement and security agencies in their criminal or national-security investigations.

1.2 The Scant Bibliography on Communications Interception

For a moment, I thought about telling the police detective about my obscure specialization in the highly esoteric subject of wiretapping. But I decided not to. Who knows, it might have made him suspicious, I thought. After all, wiretapping is not exactly a topic of conversation that one has around the dinner table. Ironically, wiretapping and its ramifications are all around us. It has been immensely impactful on both history and politics for nearly two centuries. In the words of one expert, "nations have been born and governments overthrown, thousands of lives lost and others saved, by the use of techniques and devices virtually unknown to the general public" (Spindel 1968, p. 7). Indeed, by 1998, when the telecommunications digitization and deregulation were in full swing, very few academic researchers had examined the topic. Most of them had studied communications interception practices in communist regimes around the world (Minnick 1995; Baleanu 1995), with the USSR providing a prominent theme (Ball 1989a, b; Richelson 1986; Knight 1990; Romerstein and Levchenko 1989). Other parts of the literature discussed in considerable detail the uses of communications interception made by Western intelligence services against communist regimes during the Cold War (Andrew and Gordievsky 1990; Milano 1995; Murphy et al. 1997). What was missing was research on the uses of communications interception by Western governments for purposes of internal policing—something that remains true to this day. The few exceptions came mostly from legal scholars with little technical knowledge of the subject. The principal social science disciplines, namely sociology and political science, remained largely silent.

1.3   British Literature on Analog Communications Interception

Nowhere was that absence more manifest than in regards to the British experience, where virtually no academic research had emerged on the subject of analog communications interception. One text, produced by a surveillance professional operating in the private sector, characteristically noted that "in Britain, nobody really knows how widespread the surveillance activities of government agencies are" (Wingfield 1984, p. 31). The sole remaining book on the British experience (Fitzgerald and Leopold 1987) was produced by two investigative journalists with leftwing political sympathies. Its pages were marked by politically partisan rhetoric, which remains typical of many a work on communications interception today. The book was, in effect, an attempt to address two main issues: first, to question official government data on communications interception; and second, to argue that the lack of popular oversight over the practices of the intelligence services was leading to the creation of an Orwellian state structure in Britain—a development that, according to the writers, ominously hanged in the balance:

> Britain in the eighties is not a police state in the classic, Orwellian sense. What has prevented this, however, are not the laws and administrative regulations governing tapping which pass as 'accountability' in the case of the intelligence services. Rather, our current defenses against total surveillance lie in the institutional and financial constraints put on the spies and, above all, on the restraining influence of [legal] precedent (Fitzgerald and Leopold 1987, p. 35).

The book presented yet another characteristic of many similar works on the subject of communications interception, namely an impressive body of historical research that led to mostly reductionist interpretations. According to the authors, on one side of the fence stood the reactionary forces of the conservative government of Margaret Thatcher and British Telecom, while on the other was civil society and political activists, including Labour Party militants, miner activists and trade unionists (Fitzgerald and Leopold 1987, pp. 24, 58–59, 82). Such politicized interpretations of the social and technical aspects of communications interception are typical in the relevant literature. An illuminating case in point is one of Tony Bunyan's earlier works on British intelligence institutions, entitled *History and Practice of the Political Police in Britain*. The book, written in 1976, interprets state-sponsored communications interception as being virtually synonymous with the unjustified invasion of privacy by authoritarian and fundamentally undemocratic state institutions (Bunyan 1976, various pages throughout).

On the directly opposite side of the debate one could find rare instances of British law enforcement or Home Office officials who endeavored to publicly justify the practice of communications interception. C.H. Rolph's 1973 paper entitled *The British Analogy* is largely representative of such endeavors, which are, on the whole, largely descriptive rather than interpretative. Thus, they are characterized by the absence of any political considerations of, or critical attitude toward, the practice and culture of communications interception. Instead, communications interception is considered as a customary activity of government that is not particularly different from—for instance—tax revenue or agricultural policy. He notes, characteristically:

> In 1957 [. . .a C]ommittee [of Privy Councillors] found that [communications] interception had no basis in law, whether common law, acts of Parliament or what is called royal prerogative: the only basis the Committee could discover was long usage. This is another way of saying that it is unlawful but no one cares to stop it. And long usage, though not so long as you might suppose, is the basis of a very high proportion of the powers exercised by the police. As a matter of interest, the police in England were regulating road traffic for very many years before they were given the legal power to do it by the Motor Car Act of 1903 (Rolph 1973, p. 395).

By the late 1990s, among the strictly academic examinations of communications interception, there could be found no more than three works that offered anything other than a supplementary approach to the subject within the British context. These were the works of Lambert (1986), Gill (1994) and Lustgarten and Leigh (1994). The first contains a brief section on the uses of communications interception by law enforcement, in which a crucial dichotomy is established between the police and security services. The former was believed to make limited use of communications interception, whereas the latter was thought to enjoy virtually unlimited usage of the practice (Lambert 1986, pp. 207–208). There is also an overall skepticism of officially approved figures of the use of communications interception, the latter being described as having remained at "surprisingly stable" (Lambert 1986, p. 212) levels over the years. In one characteristic passage that examines this statistical anomaly, Gill cleverly employs inference to challenge official figures:

> [o]fficial figures on the extent of tapping have remained constant for 20 years with Home Secretaries issuing between 350 and 470 warrants a year since 1969 while the number of telephone lines has increased from 8 million to over 20 million (Gill 1994, p. 168).

Though their book is not directly concerned with communications interception, Lustgarten and Leigh offer a detailed comparative approach of the legal, social and political context in which wiretapping is authorized and practiced in Britain. In a manner similar to Gill, the authors question whether there can ever be adequate executive control over security and intelligence operations in Western parliamentary democracies. But this does not mean that intelligence operations should be abolished, they say. On the contrary, intelligence and security are not alien to the roots of parliamentary-democratic governance. Democracy, like any other system of government, needs protection from its enemies, and genuinely democratic—albeit imperfect—security services can help facilitate that (Lustgarten and Leigh 1994, p. 217). However, it is the placement of intelligence institutions at the very heart of the democratic state that allows them to operate with relative detachment from the principles of transparency. Thus, when national security is perceived to be at stake, it is usually the rules of democratic accountability that are modified to accommodate intelligence requirements, as opposed to the opposite, they argue (Lustgarten and Leigh 1994, p. 24).

1.4   American Literature on Analog Communications Interception

In the United States, the pre-September 11, 2001, discussion of communications interception was deeply influenced by the country's central role in the Cold War. It is clear in the literature that, despite the political distortions that the Cold War inflicted on American political life, state agencies were far more transparent in regards to wiretapping than their British counterparts in the twentieth century. This phenomenon was molded by the traditional American mistrust of government and shaped by the polarizing experience of McCarthyism and the second Red Scare in the late 1940s and early 1950s (Wills 1999; Schrecker 1986). But perhaps the most powerful shaping force in the relevant literature is the political stench of the Watergate affair of 1972–1974.

It was soon after the downfall of Senator Joseph McCarthy that the Pennsylvania Bar Association commissioned three law professors, Samuel Dash, Richard F. Schwartz and Robert E. Knowlton, to conduct an investigation of the use and—as their probe revealed—abuse of communications interception by American law enforcement agencies. Their report, later published as a book entitled *The Eavesdroppers* (Dash et al. 1959), involved the analysis of thousands of eponymous and anonymous questionnaires, as well as hundreds of interviews with current and former police officers, wiretap practitioners, telephone company employees, district attorneys and wiretapping equipment manufacturers in dozens of American states. Meticulously detailed in the report's 500 pages are three notions that have shaped the American literature on communications interception ever since. First, there is the belief that information security, as a technical parameter of the telephone network, had been legally ignored and politically suppressed from the very genesis of the telephone system in the United States. Second, there appears to be a consistent mistrust of official government statistics on the domestic use of communications interception. Third, the authors describe the existence of an intimate institutional interface between law enforcement and the AT&T monopoly, which they view as the primary facilitator of illegal communications interception. Specifically, the technical knowledge possessed by telephone company staff is described as indispensable for the day-to-day facilitation of law enforcement communications-interception operations.

Alongside a handful of academic studies of analog communications interception, one can find a few books by practitioners who worked for the telephone industry, for law enforcement, or for both. Among them are Alan LeMond and Ron Fry, authors of *No Place to Hide: A Guide to Bugs, Wire Taps, Surveillance and Other Privacy Invasions* (1975). The two experts correctly note that, "if a local law-enforcement official wishes to initiate a wiretap with the proper court approval, dealing with the telephone company is easier than getting a dial tone. Most of the telephone companies around the country will even supply the necessary equipment to the grateful police" (LeMond and Fry 1975, p. 23). In criticizing the Federal Communications Act of 1934—to be discussed here in subsequent chapters—the authors assert that the law was deliberately designed to prohibit, not the illegal use of state-sponsored communications interception, but rather the divulgence of information gathered through such practices in court. In short, "while [non-warranted] wiretapping [is] fine, don't ever let anyone know that you've done it", they write (LeMond and Fry 1975, p. 7).

Most twentieth-century American works produced by communications-interception practitioners tend to overlap in the information they provide, as well as in their interpretations of it. An exception to the rule is the book *The Ominous Ear*, by Bernard Spindel (1968), who chastises academics and legal scholars for their "tremendous lack of understanding of the technical aspects of electronics" (Spindel 1968, p. 29). He illustrates his point using the example of a legal interpretation of wiretap law by a United States federal court, which

> declared that if the interception allowed the eavesdropper to overhear a conversation before the person for whom the communication was intended heard it, a wiretap was thereby established. On the other hand, if the eavesdropper heard the conversation even a fraction of a second after the intended recipient, there was no wiretap. Such technical splitting is difficult to accept in view of the fact that electronic sound travels at the speed of light— 186,000 miles per second. It becomes obvious that it is most unwise to delegate the task of making such decisions or laws to lawmakers who are not completely educated in the technical aspects of the field in which legislation is considered (Spindel 1968, p. 29).

It is important to point out here that the relevant literature shows virtually no awareness of the technological challenges to state-sponsored communications interception posed by the merging of electronic telephony and digital microprocessors. The technical concerns raised by Spindel in 1968 would be completely overlooked, only to resurface nearly three decades later, prompting the widespread panic among law enforcement and security agencies that this book discusses.

More political and sociological—rather than technical—interpretations of communications interception in America were provided in a number of academic works that were published in response to the Watergate Affair (Donner 1980; Theoharis 1978; Joy and Wright 1974; Navasky and Lewin 1973). The political gravity of Watergate prompted researchers to challenge the mysticism surrounding communications interception. Their works therefore tend to echo, rather than depart from, the earlier findings of Dash et al. The change of paradigm that was presented by the Watergate experience enabled these writers to employ in robust critical theory that highlighted the truly shocking evidence produced by the post-Watergate Congressional investigations of the United States Intelligence Community. In his 1978 study entitled *Spying on Americans: Political Surveillance from Hoover to the Huston*

Plan, Athan Theoharis cites a memorable passage from one such Congressional report:

> The Government, operating [...] through [...] intrusive techniques such as wiretaps, micro-phone 'bugs', surreptitious mail opening and break-ins, has swept in vast amounts of information about the personal lives, views and associations of American citizens [...]. The surveillance which we investigated was not only vastly excessive in breadth [...], but was also often conducted by illegal or improper means [. These are] domestic intelligence activities [that] threaten to undermine our democratic society and fundamentally alter its nature (USSSCSGORIA cited in Theoharis (1978), p. 110).

In light of these revelations in the immediate post-Watergate era, American academic texts on communications interception became more willing to openly dismiss official state figures of its use. Frank Donner, for example, forcefully asserts that US intelligence officials "routinely resort [. . .] to lying, deception, plausible denial, and related arts to escape detection or, on higher levels, responsibility" (Donner 1980, p. 24). Such forthright of accusations were almost completely absent from American academic writing on communications interception in the pre-Watergate era.

This drastic change of attitude in scholarly perceptions of communications interception prepared the ground for a new generation of American scholars who—though limited in numbers—managed to carry the torch of research on the subject to the threshold of the twentieth century. Two such scholars, Whitfield Diffie and Susan Landau, produced *Privacy on the Line: The Politics of Wiretapping and Encryption*, a much-needed text that somewhat bridged the crucial analytical gap between information security and politics, which has traditionally marred the subject. In their book, the two authors make full use of the hindsight afforded to them by writing after the Watergate Affair and after the Cold War. Crucially, many of the unlawful surveillance practices of American intelligence agencies had been openly publicized and acknowledged by that time. By 1998, when Diffie and Landau published their book, the topic of abuse of communications interception by American government agencies was no more a matter of speculation:

> [d]espite strictures to prevent abuses, the US government has invaded citizens' privacy many times over the last 50 years, in many different political situations, targeting individuals and political groups [. . .]. Sometimes invasion of privacy has been government policy; some-times a breach has occurred because an individual within the government misappropriated collected information (Diffie and Landau 1998, p. 148).

Diffie and Landau's analysis of the absence of information security from the telephone system is relentlessly critical. They state that "[t]he history of the last five decades shows that attacks on [telecommunications] privacy are not an anomaly. When government has the power to invade privacy, abuses occur" (Diffie and Landau 1998, p. 164).

1.5   Information Security, Bureaucracy, Secrecy and Surveillance

It is important not to confine the discussion of communications interception, as it relates to information security, within narrow technical parameters. Indeed, by its very nature, information security is a sociopolitical field insomuch as it directly informs—and is informed by—debates relating to the informational infrastructure of contemporary knowledge-based bureaucracies. It also touches on issues of transparency and secrecy, as well as on the phenomenon of surveillance, which the relevant literature tends to examine in the context of the panoptic character of administrative apparatuses in the contemporary nation-state.

Since the emergence of the work of Max Weber in the late nineteenth century, the evolution of academic research on the bureaucratic basis of the nation-state has undergone two fundamental shifts. The first shift signified a departure from earlier—primarily Marxist—interpretations of state bureaucracy as a unified bourgeois stratum that consciously subverts popular power by exercising the advantages of a technically superior ruling class (Perez-Diaz 1978, p. 78). Contemporary notions of the bureaucracy view it as a collection of multifaceted structures, which are too "polyarchic" (Giddens 1985, p. 4) in nature to systematically promote unified interests upon the citizenry (Perez-Diaz 1978, p. 77). Rather, their existence is seen as a platform of negotiation and exchange between state and civil society. This, in turn, often gives state bureaucracies a political dynamic that runs counter to the interests of both state and civil society, thus allowing bureaucrats to acquire relative autonomy from both (Lefort 1986, p. 90; Ellul 1965, p. 259; Etzioni-Halevy 1983, p. 87). The second shift centers on the introduction of the concepts of *knowledge* and *information* as vital parameters of bureaucratic functions. Weber's profound assertion that "bureaucratic administration means fundamentally the exercise of control on the basis of knowledge" (Weber 1968, p. 26) has been further-articulated by a number of scholars whose works consistently promote the definition of state bureaucracy as "the application of knowledge by political means" (Apter 1965, pp. 316–317).

The American political scientist Karl W. Deutsch was the first to make explicit references to information as the currency of political governance. In his 1963 book *The Nerves of Government: Models of Political Communication and Control*, Deutsch describes modern governments as "political systems are networks of decision and control [that] are dependent on [the] process of communications [and] the processing of information" (Deutsch 1963, p. 145). He was soon followed by others, mostly sociologists. One of them was Nicos Mouzelis, who in 1967 described bureaucratic administration as "control through knowing" (Mouzelis 1967, p. 39). In 1983 another sociologist, Eva Etzioni-Halevy, asserted that the increasing complexity of large nation-states commands that "political decisions are based increasingly on expert technical knowledge" (Etzioni-Halevy 1983, p. 58). The British sociologist Anthony Giddens elaborated on the "expansion of the 'documentary' activities" (Giddens 1985, p. 172) of nation-states throughout the twentieth century.

And writing with reference to the French experience, Michel Foucault noted that modern state administrations seek "to form a body of knowledge about individuals [...], rather than to employ the ostentatious signs of sovereignty" (Foucault 1977, p. 220).

Much of the vivid debate of the informational nature of the contemporary nation-states has taken place within the mounting scholarly literature on surveillance. The foundations of the debate were established in 1973 by James Rule's groundbreaking work, *Private Lives and Public Surveillance*. In his elaborate study of the micro-techniques of information-collection and amassment, Rule examined surveillance as a central aspect of social development. In Rule's work, surveillance was explained, arguably for the first time in the history of social analysis, not as a symptom of the bureaucratic imperatives of organized societies, but rather as one of their precondi-tions. Surveillance was thus analyzed as one of the forces that *enable* social organization:

> one may expect new means of control to emerge to ensure compliance among the partici-pants in [...] larger scale social units. In many cases it may be difficult to determine whether the growth of a larger scale social unit spurred the development of new means of control. But one may be certain that the increase in social scale will not proceed any faster than the development of means for assuring compliance within the social structures. And the development of those techniques will, where mass clienteles are involved, entail growth in the capacity of mass surveillance and control (Rule 1973, p. 308).

With the above quote, Rule accomplishes two significant achievements: first, he successfully transfers the notion of surveillance from the realm of the technical to that of the social; second, by virtue of that very transfer, he cleanses his analysis from technological-deterministic elements, which means that he does not view social models of surveillance as necessarily technological phenomena.

Rule's insights directly refute the views of scholars that are far better known than him, such as Jacques Ellul, whose techno-deterministic views on surveillance are evident in the following quote:

> [t]his entry of technology means control over all the persons involved, all the powers, all the decisions and changes [...]. Technology imposes its own law on the different social organizations, disturbing fundamentally what is thought to be permanent (Ellul 1989, pp. 134–135).

Like Ellul, Oscar Gandy proposes a view of surveillance in modern societies that is in essence technologically driven and technologically achieved. According to him, not only are the features of societies determined by their technological sophistica-tion, but society's very nature, its very guiding principles, are inherently technolog-ical. Thus, the interconnection between computerization and telecommunications networks constitutes "the material force in the new technology of surveillance" (Gandy 1989, p. 63), which "by definition [...] dramatically increase[s] the bureau-cratic advantage" (Gandy 1989, pp. 46, 62). An arguably more balanced approach is offered by David Lyon, who recognizes that the introduction of new surveillance technologies could potentially erode established civil liberties and freedoms by altering the technical balance of modern societies in favor of highly organized bureaucracies (Lyon 1994, p. 41). On the other hand, however, he goes to great pains to clarify that

> [t]he fact that information technology provides the instrument or means of this strengthened surveillance does not mean that surveillance capacity is an outcome of technological pressures. They play a part [...], but within the broader context of political, economic and cultural processes that give them their chance (Lyon 1994, p. 88).

Lyon notes that, even in the case of the intelligence services, whose operational abilities and work practices are often crucially shaped by advances in technology, the role of technology emerges through a series of social and political decisions by agents. These agents consciously wish to see national security grow dependent on new technology, he argues (Lyon 1994, p. 116). This social-constructivist view of technological development enables Lyon to walk in Rule's scholarly footsteps, especially when he insists on viewing surveillance as a multifaceted social phenomenon whose technical specifications are often decisively adapted to suit various political trends:

> the surveillance society has more than one face. It may be viewed either from the perspective of social control or from that of social participation. The administrative machinery constructed during the nineteenth century can be understood both as a negative phenomenon [...] or, more positively, as a means of ensuring that equal treatment is meted out to all citizens (Lyon 1994, p. 31).

Thus, as a technological construct, surveillance is explained in reference to a constantly evolving process of definition and redefinition, which occurs throughout history and is formed by—as well as forms—the sociopolitical landscape in which it operates. Most scholarship agrees that surveillance forms simultaneously the technical infrastructure of hierarchical repression *and* the facilitative machinery of democratic discourse. This is particularly applicable in the context of social relations in the Western liberal context. The 'rights of the individual'—an ideological cornerstone of free-market doctrines—requires that the individual and his or her needs are differentiated from those of the mass citizenship by a highly penetrative and perceptive bureaucracy. This "panoptic sort" is in essence the "all-seeing eye of the difference machine that guides the global capitalist system" (Gandy 1993, p. 1). Surveillance thus is nothing more, and nothing less, than an administrative means of reproducing the information society and its associated functions (Dandeker 1990, p. 38).

Taking this argument a step further, we can assert that the unprecedented level of bureaucratic organization that makes the information society possible has enabled an equally unmatched degree of surveillance capabilities and uses. As the infrastructure of the information society expands to encompass the entire spectrum of human activities and desires, so does surveillance. For Giddens, it is through surveillance that administrative power "increasingly enters into the minutiae of daily life [and scans even] the most intimate personal actions and relationships" (Giddens 1985, p. 309). Ultimately, the era of the information society is the age of over-penetrating bureaucracy, to such an extent that simply "to participate in modern society is to be under electronic surveillance" (Lyon 1994, p. 4; see also Gandy 1989, p. 61;

Dandeker 1990, p. 2). This statement contains something of the essence of the communications interception debate in which communications surveillance is seen as both omnipresent and inescapable—a vital aspect of governance in the information society,

1.6 A Polemical Contribution to the Literature on Information Security

I wrote this book to argue that much of the research on communications interception that was written in the analog era of telecommunications, namely before the wholesale introduction of digital switches and fiber optic networks, must be radically re-evaluated in light of digitization. I am able to do this because, starting in the autumn of 2000, my professional position afforded me the opportunity to form connections with some of the leading protagonists involved in negotiations over the digitization of telecommunications and attempts by the governments of the United States and Britain to secure their ability to wiretap. As I traveled regularly between Britain and the United States, I closely monitored these conversations until their conclusion in the lead-up to the dramatic events of 11 September 2001. I spoke with members of the security services, law enforcement agencies and other government bodies in the United States and the United Kingdom. I also spoke with security officers working for large British and American telecommunications service providers, as well as with senior officials of telecommunications industry bodies who represented hundreds of companies in negotiations with government officials. Finally, I discussed these issues with civil-liberties advocates, who voiced the views of the organized civil-liberties lobby in the negotiations between the telecommunications industry and the government.

My strong interest in the information-security ramifications of these discussions prompted me to have prolonged conversations with these individuals. These were consensual and recorded, and I was provided with permission by the participants to make use of them for academic research. The only provision requested by the participants in the conversations was that they remained anonymous, due to the political sensitivity of discussing issues that had a major impact on national security. In 2001, as a direct result of the history-altering events of September 11, 2001, my research interests changed radically, as I entered the then-emerging discipline of intelligence studies, where I remain to this day. Consequently, the recordings of these conversations were set aside and never used for publication. But in the past couple of years, as the negotiations between government and industry on the topic of communications interception approached their 20-year mark, I decided to revisit the data and provide a historical assessment of that critical juncture in the evolution of modern-day information security. Now, nearly two decades after the conclusion of these prolonged negotiations, my face-to-face access to these negotiators permits me to present a detailed re-evaluation of that period that crucially shaped the current

state of information security in the area of telephony. This qualitative methodological approach provides a context for the decisions that were taken and allows for a
qualitative analysis of the outcome of the negotiation process. More importantly,
given the veil of official secrecy that surrounds the subject, this book offers the only
publicly available glimpse into the minds of negotiators from all sides of this intense
debate.

It is indeed interesting to observe that, in the ensuing two decades, the topic of
wiretapping remained completely absent from academic scholarship, with a handful
of commendable exceptions (such as Landau 2010). Even those notable exceptions,
however, do not take into consideration the views of the actual people who were
involved in the tense and politically consequential negotiations over communications interception in the late 1990s. This book, therefore, provides the first-ever
comprehensive analysis of the evolution of wiretapping from the beginnings of
telephony until the dawn of the twenty-first century. In doing so, it focuses on that
unique period in the late 1980s and 1990s, when state agencies faced—for the first
time since the emergence of telecommunication—the very real prospect of losing
their ability to intercept.

In formulating its argument, this work follows in the footsteps of much of the
literature on the topic of telecommunications, which refuses to examine the evolution of information security in isolation from its social and political aspects. It echoes
Rule's argument that the models of information security, on which our social order is
based, were developed through a complex interplay of sociotechnical factors. These
factors were crucially informed by political—and not technical—decisions (Rule
1973, p. 308). In recognizing this fact, one must not be naïve. As this book will
demonstrate, the history of wiretapping shows that researchers have been right to
question official government data of its use by state agencies. The faith that
advanced Western societies place on technology has contributed to the development
of policing, security and intelligence models that rely enormously on communications technology. Consequently, communications interception has been probably the
most constant technological feature of Western security apparatuses over the past
century. This dependence—some, e.g., Macrakis 2010 (various pages throughout)
would say over-reliance—on communications interception has been especially
noticeable during existing or perceived political crises. Such crises, for instance,
were the second Red Scare or the Vietnam War in the United States, or the Troubles
in the United Kingdom. In those times, when the use of communications interception
by state agencies assumed extra-legal characteristics, its excesses were facilitated by
the intimate institutional interface between state agencies and the telecommunications service providers [Dash et al. 1959 (various pages throughout); LeMond and
Fry 1975 (p. 23); Lambert 1986 (207 ff)]. During such periods of crisis, the
information security features of telecommunications networks was systematically
subverted by both state agencies and service providers, often with little regard to
technical constraints, legal parameters or ethical doctrines.

This leads us to the assertion that the view of authors like Fitzgerald and Leopold
that society's defenses against total surveillance rest on "institutional and financial
constraints [and] on the restraining influence of [legal] precedent" (Fitzgerald and

Leopold 1987, p. 35) is wrong. As this book will show with reference to the British and American telecommunications experience, institutional constraints and legal codes have historically been unable to limit privacy abuses by the government. At the same time, it is also wrong to assert that the technological evolution of telecommunications is inevitably "altering the technical balance of modern societies in favor of highly organized bureaucracies" (Lyon 1994, p. 41). The case study examined here demonstrates that the "material force" of technological change does not "by definition [...] dramatically increase the bureaucratic advantage" of the state (Gandy 1989, pp. 46, 62). In fact, technological change advances the surveillance apparatus of the state only when the latter is in control of the material forces of production that birth these technologies. When these forces operate outside direct state control—for instance in the private sector—the surveillance needs of the state are not necessarily taken into consideration during the process of technological innovation.

In the case of the digitization of telecommunications, not only were the state's security requirements an afterthought, but the state itself was caught completely unaware of the groundbreaking speed and immense power of technical innovation that was taking place in the private sector. In the words of Spindel, state agencies that relied heavily on communications interception for their day-to-day operation showed a "tremendous lack of understanding of the technical aspects of electronics" (Spindel 1968, p. 29) in the area of telecommunications. In essence, the case study examined in this book contests commonplace views of the state bureaucracy as a proactive depository of "expert technical knowledge" (Etzioni-Halevy 1983, p. 58). In the case of telecommunications, the transfer of technical know-how from the government sector to private industry, which was facilitated by the deregulation of the 1980s, effectively challenged the state's ability to "exercise control on the basis of knowledge" (Weber 1968, p. 26; see also Mouzelis 1967, p. 39).

This is not to say that the information-security principles of telecommunications networks are not susceptible to regulatory—i.e. political—pressure from government agencies. This book will demonstrate that when its ability to exercise bureaucratic control through policing is challenged, the state will attempt to recover its policing abilities by any means necessary, including aggressive regulation and even nationalization. In the context of the digitization of telecommunications, the British and American governments chose to actively suppress the technical features of the network that promoted information security. They did so in order to protect the ability of policing and intelligence agencies to intercept the pattern and content of communications activity. But that process was neither simple nor easy; in fact it has yet to be fully implemented. Furthermore, it demonstrated that the entry of technology does not mean "control over all the persons involved, all the powers, all the decisions and changes" (Ellul 1989, pp. 134–135), nor does it inevitably lead to "the creation of an Orwellian state structure" (Fitzgerald and Leopold 1987). On the contrary, rapid technological development, in combination with privatization and deregulation, subverted the state's ability to exercise surveillance and control on the much-heralded information superhighway.

Ultimately, the modern state apparatus—in the United States, Britain, and elsewhere—is anything but monolithic. It is neither authoritarian and fundamentally undemocratic (Bunyan 1976; various pages throughout), nor is it the absolute guarantor of individual and collective rights in democracies. Rather, it is complex and often self-contradictory. It is, as Giddens states, "polyarchic" (Giddens 1985, p. 4) and does not systematically promote a set of unified interests (Perez-Diaz 1978, p. 77). The same truism applies to civil society in our time, whose relationship with information security as a technical and social principle is complex and ambivalent at best. The case study of telecommunications digitization, which is examined in detail in this book, makes it clear that it is difficult to separate freedom from surveillance. In other words, it is difficult to distinguish those technologies that promote individualism and freedom of action, which are abundantly afforded by the information society, from those technologies that subvert privacy. The same telecommunications features that facilitate online voting in a fully or partly digitized system of democratic participation also allow state agencies to direct their formidable surveillance arsenal against individual targets. In the words of Lyon, "to participate in modern society is to be under electronic surveillance" (Lyon 1994, p. 4). In the information society, therefore the line that separates the machinery of democratic discourse from the technical infrastructure of Jeremy Bentham's *panopticon* is paper-thin. In the case of communications interception, we can state with certainty that the decision to subvert the information-security features of the telecommunications network was a political choice, not a product of technical inevitability. As Lyon stated in his book *The Electronic Eye: The Rise of Surveillance Society*, the technical specifications of surveillance were "decisively adapted to suit a variety of political trends" (Lyon 1994, p. 31). We conclude, therefore, that information security is not a technical, but rather a sociopolitical, concept. It is technology that embodies a host of—often competing—political, economic, social and cultural processes. These will be explained in detail in the ensuing chapters of this book.

Chapter 2
The Context

The primary purpose of every communication system is to convey intended meanings from a sender to a receiver through the transmission of information. When the transmission of information is mediated through the use of technology, it becomes *telecommunication*—namely the transmission of information through the use of optical, electromagnetic, or digital systems of communication. It stands to reason that the principal objective of every telecommunication system is to be *effective*— that is, to transmit information efficiently.

Additional complications, however, appear when political parameters inform the communication process. An example is when the content of a message is meant to be communicated solely to an intended receiver. Another complication is when a third party, who is not the intended receiver of the information, attempts to access it by overt or clandestine means. It is complications such as these that form the basis of the practice of *information security*, whose purpose is to prevent unauthorized access to confidential information while maintaining the efficiency of the communication exchange.

Unsurprisingly, the historical development of information security in the past two centuries has progressed in parallel with the development of telecommunications. From its very inception in the 1830s, the electrical telegraph—history's first large-scale electrical telecommunication system—raised concerns among its users about potential security breaches (Phalen 2015, p. 185). Such concerns became central as the telegraph was utilized in war—notably in the Crimean War (1854–1856), where it was first tested in interstate combat, and in the American Civil War (1861–1865) (Sterling et al. 2007 (142ff)).

As electrical telecommunications continued to evolve, organized governments began to view them as facilitators of economic, social and political order (Shannon 1958, 1959; Lipset 1959). This was later noted by Professor Wilbur Schramm, one of the founders of communication studies, who said that

> [w]ithout a sophisticated and efficient development of communication, the base of population, co-operation, industrialization, education and skills needed in a modern industrial society cannot possibly be established (Schramm 1964, p. 41).

© Springer Nature Switzerland AG 2020
J. Fitsanakis, *Redesigning Wiretapping*, History of Information Security,
https://doi.org/10.1007/978-3-030-39919-1_2

By the second half of the nineteenth century, as the empires of the modern era grew to encompass territory on every continent, efficient telecommunications systems became indispensable for rational administration and planning across large distances (Spengler 1963, p. 211).

2.1   Regulating and Policing Telecommunications Networks

At the same time, however, organized states began to use telecommunications systems to facilitate the management and regulation of public life, as well as to deter threats to their integrity and—ultimately—their ability to exercise power (Gill 1994, p. 55; Spitzer 1985, p. 325). That was largely done through policing, which took the form of state-sponsored *telecommunications surveillance*. The latter can be defined as the ability of state authorities to access and comprehend the pattern, as well as the content, of telecommunications activity. Historically, state-sponsored surveillance has been applied to domestic telecommunications networks in order to help combat crime, prevent political violence and—regrettably—police the political views of targets. When practiced on international telecommunications networks, state-sponsored surveillance has traditionally aimed to collect intelligence information from foreign state or non-state actors in order to satisfy national-security requirements.

As with any issue concerning civil liberties, the practice of communications interception has always been controversial; however, the secrecy that has historically surrounded the practice has prevented it from being politically contested in any significant way. By the 1890s, when telephony began to emerge as a viable method of long-distance communication, modern governments had already developed sophisticated methods of telecommunications surveillance. These were based on state prerogatives that went as far back as the emergence of the first long-distance mail services in the sixteenth century (Johnson 1998, p. 12). These prerogatives were "deduced from history and enacted into law" (Pierce 1977, p. 193), and were rarely, if ever, challenged in the political sphere.

The growth of telephony, which by the mid-twentieth century had largely displaced electric telegraphy, gave rise to the phenomenon of what is commonly referred to as *wiretapping*. The term describes the act of intercepting messages exchanged through the telephone network, usually through a main switch post (Carr 1998, Sect. 1.3). Wiretapping is not to be confused with *bugging*, which suggests audio surveillance through concealed microphones that are not connected to the telephone network. In most cases, the legal sanctity of telephonic wiretapping—or, if we are to use its accurate technical term, *communications interception*—was inherited from the age-old practices of state-sponsored mail and oral interception. Specifically, therefore, the phenomenon of state-sponsored telephonic communications interception signifies the transference of the already established principles of mail and oral interception into the domain of electric and—eventually—electronic communication.

For most of its existence, the phenomenon of communications interception has taken place within the confines of centralized states. That has certainly been the case in the United Kingdom, where the government controlled the telephone network for most of its existence. That was not always the case. In fact, telephony in Britain was developed by private entrepreneurs until 1912, when the British government nationalized the telephone network. That was a strikingly paradoxical move, given that Britain was at that time a major hub of industrialism and the epicenter of global capitalism. The government, therefore, was not favorable to enforced nationalization—a practice that was soon to become synonymous with the emergence of Bolshevism in Russia. But, as this book will show, the British state recognized that, like the military and the police, telecommunications were too important for national security to be left to the private sector. It is important to keep in mind that the nationalization model of telephony was followed by most developed countries at the time (Levi-Faur 2002, p. 163). A rare exception to that norm was telephony in the United States, which did not evolve through nationalization, but through fierce competition in the private sector. However, as this book will show, the American government exercised strict regulatory pressure on the private telecommunications sector and went out of its way to ensure that its security and law enforcement agencies had unrestricted access to the network so as to police it.

2.2 The State and Telecommunications Providers

This book will also demonstrate that the intimate institutional interface between the state and the telephone service providers—whether publicly or privately owned—facilitated the normalization of state-sponsored wiretapping for most of the twentieth century. Throughout that time, there were instances of *extralegal wiretapping*, where no regulatory framework existed to govern the practice domestically. There were also instances of *legally sanctioned wiretapping*, where the government followed existing laws that regulated the communications interception in the domestic sphere. Finally, there were instances of *illegal wiretapping*, in which government agencies bent or outright disregarded the law in order to wiretap domestic targets. In the United States, illegal wiretapping was especially frequent—some say routine—during the Cold War, when the view that the country faced an existential threat in the form of communism prompted many to disregard established civil liberties. Excesses of that period—outlined in detail in this book—involved the use of state-sponsored wiretaps against leaders of the civil rights movement and against anti-Vietnam War protestors. In the United Kingdom, where the pressures of the Cold War were arguably less intense, the security services were generally less prone to violating the law; however, the decades-long domestic unrest in Northern Ireland stretched the practice of wiretapping to its legal limits. One way or another, however, state-sponsored communications interception was practiced unabated in both countries throughout the twentieth century.

At the beginning of the 1980s, however, things started to shift drastically and unexpectedly, due to a perfect storm of major economic, technological and cultural shifts. These shifts caught many governments by surprise and directly threatened the ability of state agencies to intercept domestic communications. The purpose of this book is to document that unique and unprecedented moment in the modern history of information security, a period of time when governments found themselves temporarily unable to wiretap.

2.3 The Onset of Digitization and Informatization

What exactly happened? By the mid-1970s, it was becoming clear that telecommunications systems were not simply enablers of social and political order; they could in fact become facilitators of substantial—indeed extraordinary—economic growth. That awareness emerged more or less in parallel with the onset of *digitization*, which is the process of converting analog information into binary digits, thus making it computer-readable. Digitization facilitated the replication of information because digital data could be reproduced much faster than analog information. It also facilitated increased communication speeds because binary signals could be transmitted faster than analog signals. Finally—not necessarily in order of importance—digitization decreased the cost of storage, since digital data took much less physical space than analog data. The process of digitization—popularized through the emergence of digital computer systems—gave rise to the concept of "the information society', a phrase coined in 1977 by the French scholars Simon Nora and Alain Minc (1978). Implicit in the term was that the emerging information technologies signified a new mode of technological reasoning, which was quickly becoming the primary contributor to a new era of economic development in advanced post-industrial societies. Proponents information technologies maintained that their emergence—a trend known as *informatization*—represented a new *historical paradigm*. They saw informatization as a period of major historical change characterized by a new mode of development, described as "the informational mode of development" (Castells 1989, p. 12).

The information society model consisted of three central elements: the technological, the financial, and the societal (Miles 1988, pp. 7–8). The technological element attested to the primarily technical differences between the industrial and information societies; these differences included both the shrinking cost of information processing (Russell-Neuman 1991, p. 4; Woods 1993, p. 95) and the shrinking size of digital storage and processing equipment through advances in computing (King 1984, p. 3; Heap et al. 1995, p. 93; Woods 1993, p. 110). The financial element addressed the economic regeneration caused by the informatization of consumer capitalism, and this included the falling cost of digital communications (The Economist 1995, p. 5; Dordick and Wang 1993, p. 26), as well as the introduction of new consumer services and investment opportunities (Department of Trade and Industry 1996, p. 7; European Commission 1995, p. 13). It also included the dynamic transformation of production systems facilitated by advanced digital techniques (Miles 1988, p. 9; Dordick and Wang 1993, p. 124). The latter trend rendered corporate structures "information-dependent" (Cruise-O'Brien and Helleiner 1983, p. 3), a phenomenon through which informatization gradually became "the core, fundamental activity conditioning the effectiveness and productivity of all processes of production, distribution, consumption and management" (Castells 1989, p. 18). Thus, digital information mechanisms operating in an information society began to be viewed as the major—if not the primary—determinants of economic growth in post-industrial societies. They were also viewed as core elements of the predominant mode of production, to the extent that one could not adequately address economic development without reference to informatization. Finally, the societal element in the information society theory attested to the quantitative and qualitative change of the networks of telecommunication, through which social interaction increasingly took place and assumed meaning. In other words, as more and more areas of social activity were mediated by digital communications (Poster 1990, p. 1), the digital infrastructure of the information age gradually became proportional to the entire spectrum of human activities and desires (European Commission 1995, p. 20). In essence, digital information systems did not represent a new form of social discourse imposed upon an older one, but rather a complex weaving of different means of expression within that same discourse.

2.4 Deregulation of the Telecommunications Sector

All that is to say that, by the early 1980s it was difficult to conceive of a model of advanced socioeconomic growth that did not include rapidly accelerating versions of telephones, televisions and—eventually—networked computers. Moreover, the information society concept came with a distinct sense of openness, promises of information-sharing and non-hierarchical structures. The latter were notably reflected in the architecture of the first regional computer networks that gradually gave rise to the Internet. For many, these visions embodied the sharp contrast between the democratic West and the oppressive totalitarianism of the Soviet-led Eastern Bloc. Among the most powerful proponents of the information society vision were two seminal political figures of the 1980s, United States President Ronald Reagan and British Prime Minister Margaret Thatcher (Demac 1988, 140ff; Lean 2016, 88ff). Importantly, their vision of the information society was intimately intertwined with their strong ideological conviction in *deregulation* and *privatization*—that is, the view that government ownership and interference in the economy was inherently obstructive and hindered the dynamism of capitalist growth. Unsurprisingly, therefore, President Reagan and Prime Minister Thatcher were ideologically opposed to government ownership of a large array of service sectors, even those that were traditionally considered essential. Explaining her view on the topic, Thatcher said that privatization was fundamental to improving Britain's economic performance. But for me it was also far more than that: it was one of the central means of reversing the corrosive and corrupting effects of socialism. Ownership by the state is just that—ownership by an impersonal legal entity: it amounts to control by politicians and civil servants [...]. Just as nationalization was at the heart of the collectivist programme by which [previous leftwing] governments sought to remodel British society, so privatization is at the center of any programme of reclaiming territory for freedom (quoted in Evans 2004, p. 27).

In 1979, when Thatcher became prime minister, Britain was closer to the European postwar model of economic development. It therefore subscribed to the basic tenets of capitalism, but—unlike America—integrated it with a strong emphasis on a comprehensive system of social welfare. The government's involvement in industry and essential services, such as banking, housing provision, telecommunications and transportation, was substantial. By the time Thatcher retired from politics, in 1990, the British government's role in the economy had been drastically curtailed (Evans 2004, p. 12) and its function resembled the American model of minimalist government far more than previously.

Throughout his two-term presidency, Ronald Reagan fought to further-reduce government regulation, stating in his inauguration speech that "government is not a solution to our problem; government *is* the problem" (Reagan 1989, p. 61). A major test-case for Reagan's views on government regulation centered on one of America's most powerful conglomerates, the American Telephone and Telegraph Company, known as AT&T. Having emerged in the 1880s during the dawn of commercial telephony, AT&T and its subsidiary companies operated under a peculiar state-authorized monopoly that allowed it to become arguably the world's most powerful telephone company. That monopolistic arrangement, referred to as the *Bell System*, stemmed from the government's view that an essential public service like the telephone would benefit from the economic integration of the telecommunications network and its control by a single company. Consequently, as it is often put by historians, an AT&T monopoly was "not discouraged" by government (Canes 1966, p. 12), which saw telephony as an important national asset. But Reagan disturbed that century-long arrangement. His administration saw the breakup of the AT&T monopoly as "the key not only to opening up innovation but also to deregulating telecommunications" as a whole (Hayward 2009, p. 216), in pursuit of the information-society vision. Consequently, by 1984, a series of lawsuits brought by the government had broken up AT&T into seven separate regional companies, a move that almost immediately sparked the unprecedented proliferation of telecommunications service providers (TSPs) in the world's largest economy.

In Britain, the national telephone network had developed under direct state ownership since 1912, when—as explained earlier—the government simply took over the privately owned National Telephone Company (Perry 1977, p. 90). At that time, the telephone system was brought under the control of the General Post Office, which had operated as the British government's communication arm since 1660. In 1969, while remaining under direct state control, the telephony division of the General Post Office was renamed Post Office Telecommunications. But in 1981, the Thatcher government renamed the state-owned company to British Telecom. The following summer the government announced that it would begin to offer shares in British Telecom for sale to the public. By 1984, the year when AT&T was officially dismantled in America, British Telecom had been incorporated into a public limited company and its shares were being publicly sold and traded (Marino 2007, 114ff).

Consequently, by 1984, the Reagan administration in the United States and the Thatcher government in Britain had radically disrupted the institutional arrangements of their countries' respective telecommunications sectors. Two companies, AT&T and Post Office Telecommunications, which had enjoyed unquestioned government protection for a nearly a century, were suddenly forced to compete in a new and deregulated environment. Proponents of limited government and deregulation rightfully rejoiced in seeing such seismic changes on both sides of the Atlantic. The conservative revolution of the 1980s was well underway.

2.5 Privatization and Deregulation as Anathemas Wiretapping

In the midst of all the celebrations, however, two closely linked sectors of government were teetering on the edge of the abyss. British and American law enforcement and security agencies were finding that they were becoming increasingly unable to carry out communications interception. The reasons were multifold, but primarily stemmed from the deregulation and privatization revolution of the 1980s. Unquestionably, the privatization of the telecommunications landscape had helped generate the mushrooming of a vast new service and manufacturing sector, which consisted of literally thousands of new, privately owned companies. Thus, the institutional landscape of telecommunications was widening with unprecedented speed. As a result, the institutional interface between the security agencies and the telecommunications sector was becoming vastly more complex. Before deregulation, the government agencies that were in pursuit of wiretapping mandates were used to dealing with a single point of contact in a single company: AT&T in America, or Post Office Telecommunications in Britain. They were now finding that different parts of the telephone network were owned by a milieu of different companies. Moreover, the technical infrastructure of telecommunications had fragmented: in some cases, the companies that owned the phone lines were different from the companies that provided communications services to consumers. And there were external third parties that manufactured, owned or operated the switches that carried or amplified voice messages. These new companies were national, regional, or even local, with differing sizes, cultures, levels of experience and—most of all—differing degrees of willingness to assist law enforcement and security agencies in their criminal or national-security investigations. That new and unprecedented situation marked a stark contrast to what used to be a unified telecommunications network.

To make things worse, the unprecedented technological growth that had been unleashed in the new, privatized telecommunications landscape had given birth to a host of new digital services and methods of information exchange. Digital telephone features such as speed-dialing, call-forwarding, remote answering services and voicemail, teleconferencing, number portability, and many other added services, flourished through the competition between telephone service providers. These added services posed increasingly serious technical barriers to the government's communications-interception capabilities. One Federal Bureau of Investigation (FBI) special agent recalls that, in the late 1980s, the FBI

> became concerned about this when we observed these long periods of inactivity on the telephone when we knew [the targeted individual] was talking to someone on the telephone line. We could either observe it or we had [. . .] a microphone in there or an informant. We knew that he was using the telephone but for some reason our equipment was not working and not picking this up. I guess it first came to my notice when we had prisoners [. . .] in a local penitentiary [...] making calls to an associate only to be conferenced by that associate to a third party. And the associate would then drop out of the conversation or [. . .] would hang up his phone [. . .]. Of course, the wiretap would be on the associate outside of the prison. And we would be left high and dry with no content of the conversation, while all of the time knowing that he's talking to a drug associate or something and that the conversation was being supported by these associate's facilities and services at the phone company. We would [. . .] certainly get the billing record of it, but we had no idea what was said (FBI Special Agent and Telecommunications Industry Liaison, CALEA Implementation Section, Interview 40 2000, pp. 30–38).

By 1991, it had become obvious that that the technical specifications of the telecommunications revolution were overpowering traditional models of state-sponsored wiretapping. As this book will explain, the binary transformation of oral messages, the packetization of digital data, as well as the introduction of digital switches and digitally-enhanced wireline and wireless user services, threatened state-sponsored communications interception with virtual extinction (Yarbrough 1999, pp. 4–5; Boucher and Edwards 1994; Morris 1998). For the first time since the emergence of telegraphy in the 1830s, government agencies in the United States and the United Kingdom were finding themselves technically incapable of intercepting information exchanged on telecommunications networks. They were facing a rapidly growing technological blind spot that was denying them access to what they saw as one of the most critical techniques in their investigative and intelligence-collection arsenals.

2.6   The Introduction of Wiretapping Legislation

The situation was literally unprecedented, and prompted immediate action from government agencies on both sides of the Atlantic as security and law enforcement agencies tried to contain these new threats to their interception capabilities. As a result of intense consultations with policy-makers, two extraordinary pieces of legislation emerged. On 25 October 1994, the United States Congress passed the Communications Assistance for Law Enforcement Act (CALEA), also known as the Digital Telephony Act. Six years later, on 26 July 2000, the British Parliament

passed the Regulation of Investigatory Powers Act (RIPA). Both pieces of legislation, which have remained in force ever since, required that TSPs redesign their
existing networks to ensure that the government would be able retain its wiretapping
capabilities. They also stipulated that all new telecommunications networks and
services be built with wiretapping-enabling features to accommodate the government's wiretapping mandates.

As the world approached the year 2000, it was indeed difficult to underestimate
the technical significance of CALEA and RIPA. Their political controversy was
equally inescapable. On the one side of the argument, governments were alarmed by
the prospect of losing the ability to carry out comprehensive telecommunications
surveillance, something that had been seen as their prerogative since the 1830s. In
the United Kingdom, the National Crime Intelligence Service (NCIS) warned that, in
the absence of RIPA, the government's interception capability would "move to
redundancy" (HCTISC 1999a). In the United States, the FBI—a leading proponent
of CALEA—described the problem of digital communications interception as "one
of the most difficult [and] complex [. . .] ever to confront law enforcement" (Freeh
1997). However, critics of CALEA and RIPA were equally alarmed by their political
significance. It was noted that the two laws marked the first instance in the history of
Western telecommunications where virtually all service providers were required to
modify the technology of their systems in accordance with the requirements of law
enforcement and security agencies (Berman and Dempsey 1996). It was "the first
time in our history', as the American Civil Liberties Union (ACLU) put it, "that an
entire industry [has been] required to alter its technology so the government would
be guaranteed success in its snooping" (Steinhardt 1995). As can be expected,
consensus between government and industry bodies over the scope and scale of
CALEA and RIPA proved particularly elusive. The sheer financial scale of the
required implementations—which included the manual modification of millions of
digital switches—alarmed industry representatives. At the same time, American and
British civil-liberties watchdogs warned of the serious legal ramifications of
hardwiring interception capabilities into telecommunications networks.

2.7   Implications for Information Security

In addition to the historical analysis of CALEA and RIPA, the pages that follow offer
projections into the future of information security, which will be highly shaped by
the evolution of wiretapping legislation. In America, a country with a widespread
litigious culture, the 25-year-long CALEA debate has been decisively shaped by
numerous court cases involving the FBI, Department of Justice, as well as various
telecommunications industry associations and civil liberties groups. In the United
Kingdom, the vague language that characterized many of RIPA's early mandates
caused widespread skepticism among the industry and civil liberties advocates
(Sutter 2001). Consequently, it took years for the technical standards of the legislation to emerge, and to a certain degree they continue to evolve today, nearly 20 years

after RIPA was first passed by Parliament. CALEA too remains in flux, as new technical features—such as smartphones, strong encryption or peer-to-peer networks, to name a few—continue to inform the ongoing digital telecommunications revolution.

Therefore, as the political struggle between technology and legislation remains fierce, it is exceedingly difficult to forecast the future of information security in our century. Yet CALEA and RIPA remain as the point of origin of any informed analysis of the shape of things to come. Their emergence has been of overwhelming significance for research, inasmuch as they represent the first attempts by American and British governments to define the legal, technical and administrative parameters of communications interception in the digital environment. They will therefore be highly instrumental in shaping emerging models of state-sponsored surveillance in the information society. Their precise impact, however, remains unclear. In an important sense, the still-evolving state of CALEA and RIPA is due to the technical and political complexity of the communications-interception debate. By simply reading the text of the two pieces of legislation, it is easy to interpret them as inflexible commands issued by security-oriented policy-makers to the telecommunications industry. Yet reality is much more convoluted. Both CALEA and RIPA represented attempts to strike a balance between information security and national security—to ensure individual liberties and facilitate open, consumer-oriented markets, while fulfilling policing and national-security mandates. What is more, since the 1980s, the communications interception debate in the West has occurred within the context of increasing market deregulation at the local, national and international levels. Deregulation has the tendency to marginalize state objectives—including national security objectives—in favor of corporate and consumer goals. It is within this very context that the concept of communications interception is discussed in the present book, which endeavors to examine communications interception as a feature of information security in our digital era.

Chapter 3
The Government and Telecommunications: A Complex History

On the morning of 7 March 1876, the United States Patent Office issued the Scottish-born inventor Alexander Graham Bell a copyright license for an apparatus that would eventually come to be known as the *telephone*. The license for the new contraption was reportedly issued less than 2 h before another inventor, Elisha Gray of Western Union—then the dominant telegraphy company in America—would have been granted a patent for a very similar device (Martin 1991, p. 1). This striking account is indicative of the early development of the telephone. Its evolution was essentially the outcome of a series of historical accidents and a complex combination of technical, commercial, political and ideological interests expressed—often aggressively, other times more subtly—by a variety of actors.

This chapter illuminates some of the central elements of that complex evolution with particular reference to the relationship between telephony and the government. In the interests of coherence, the chapter is partitioned into two main sections. The first section is concerned with the early attitudes of governmental bodies and agencies in the United Kingdom over the emergence of the telephone network in the British territories. In its present state, the literature on the subject appears to be somewhat reductionist and focuses almost exclusively on the financial parameters of that reaction. The purpose of the present chapter, therefore, is to further our current historical understanding of the state's reaction to the new technology, by introducing a number of social and political parameters to the debate. The second section is concerned with the United States and discusses the early attitudes of the government toward the appearance of telephony in that vast nation. This latter section will be rather briefer, as the historical understanding of the subject in the relevant literature appears to be more conscious of the social and political elements that informed governmental decisions and actions.

© Springer Nature Switzerland AG 2020
J. Fitsanakis, *Redesigning Wiretapping*, History of Information Security,
https://doi.org/10.1007/978-3-030-39919-1_3

3.1   Messenger Boys and Telephones

Ironically, Bell found it considerably easier to develop his telephone than to convince governments around the world that it was worth taking an interest in. Senior civil servants in Britain proved to be a particularly unyielding lot. In September of 1877, after attending one of Bell's public lectures about his new invention, the British General Post Office's engineer-in-chief was not impressed. He wrote an official report against the idea of adopting the telephone, stating that the potential of its practical use in Britain and the Empire was extremely limited (Robertson 1947, p. 11). Equally unimpressed by Bell's presentation was the General Post Office's chief civil servant, who made his opinions known in several reports to his superiors in the British government (Perry 1977, p. 72). The General Post Office thus repeatedly turned down the opportunity to purchase the rights to Bell's British patent (Aronson 1977, p. 16). The latter was subsequently acquired by a consortium of private entrepreneurs who in 1878 sponsored the construction and operation of the nation's first telephone network in London. On 14 June 1878, these entrepreneurs, acting under the instructions of the Bell Telephone Company, established the first telephone firm ever to operate in Britain, the Telephone Company Ltd. The company, which was later renamed National Telephone Company, Ltd., was officially registered with a capital of £100,000 (Robertson 1947, p. 12).

The operation of the country's first telephone network had little, if any, effect on the highly negative attitudes on telephony among senior civil servants: in 1879, the new Engineer-in-Chief of the General Post Office, Sir William H. Preece, reported to the House of Commons that he saw no point in further-extending the existing telephone network. In one characteristic passage if his report, Preece justifies his perception of the telephone as a useless contraption by resorting to that omnipresent feature of British life, namely social class:

> I fancy the descriptions we get of [the telephone's] use in America are a little exaggerated, though there are conditions in America which necessitate the use of such instruments more than here. Here we have a super-abundance of messengers, errand boys and things of that kind [...]. The absence of servants has compelled Americans to adopt communication systems for domestic purposes. Few have worked at the telephone much more than I have. I have one in my office, but more for show. If I want to send a message, I use a sounder or employ a boy to take it (quoted in Dilts 1914, p. 11).

From our contemporary hindsight, the shortsightedness of Sir William's views is indeed amusing. But his stubborn refusal to secure the exclusive rights to the manufacture and operation of telephones on behalf of the General Post Office represented the dominant view among the government's senior civil servants. However, his aloof attitude was by no means shared by middle-rank administrators and engineers at the General Post Office. A careful examination of the surviving records of debates about the telephone reveals the existence of an influential group of government policy-makers who fiercely rejected the notion that a British telephone system should be allowed to develop outside the boundaries of tight state control (Perry 1977, p. 83, 1992, 149ff). This group of policy-makers, electrical engineers, telegraphers and other technical managers, proceeded to eagerly assess the emerging dynamics of telephony practically under the noses of their unenlightened superiors. Historian Charles Perry describes the seriousness with which the General Post Office took to the telephone: "By March 1877"—which was before Bell approached he British government to market his new contraption—the General Post Office had already commissioned a "report on the capabilities and practical utility of the telephone', he writes. A subsequent set of studies was "made of telephone technology in America', funded by the General Post Office (Perry 1977, pp. 72–73).

The seriousness with which the General Post Office took Bell's invention was clearly reflected in its decision to assume the crucial role of agent of the privately owned Telephone Company, Ltd, in return for receiving 40% of the company's gross rental income (Perry 1977, p. 73). Indeed, a convincing argument can be made that the General Post Office's initial decision to abstain from direct control over the new technology was anything but an outcome of ignorance or heedlessness. Rather, the British government—already in control of the nationwide telegraph and postal networks—preferred to lie in wait and allow private investors to attempt to overcome a host of early impediments to the development of the telephone. Such impediments included the problem of voice amplification over distances, not to mention geophysical threats to transmission cables. In the absence of a well-funded engineering effort of considerable magnitude, these problems could have seriously hindered—even terminated—the future of telephony. Even as late as June of 1890, as *The Times* and *The Economist* were frequently running editorials urging government ownership of the telephone system, the Treasury Department's Senior Civil Servant, Reginald Bundle, ruled out any such possibility. He insisted in Parliament that government executives were "not prepared to embark upon another enterprise gigantic in itself, while the developments it might lead to are beyond their powers of prediction" (quoted in Perry 1977, p. 88). It appears clear, however—particularly when examined in light of its subsequent course of action—that the guiding premise of the British government's early policy toward the telephone was one of patiently observing the emergence and initial development of the enterprise, "without incurring more than a minimum of direct responsibility" (Robertson 1947, p. 46).

3.2   Telephony's Takeover in the United Kingdom

In the late nineteenth century, the principal legislative safeguard that enabled the General Post Office to distance itself from early developments in telephony was to be found in the legal precedents of mail and telegraphy nationalization in Britain. According to these, the British government could assume direct and overall control of the telephone system at any point in time, and for whatever reason it saw fit, without being required to seek the industry's approval or indeed permission. The following extract from the minutes of a Treasury Department meeting on 8 April 1899, is indicative of the legislative assurance with which government officials viewed the development of telephony:

[t]he right of the Post Office to establish telephone exchanges and to license persons, companies and other bodies to establish such exchanges (a right explicitly reserved in every license and agreement with the National Telephone Company) will remain intact; and should Her Majesty's government at any time see fit to exercise this right in a manner different from that indicated in this minute, neither the company nor any local authority will have any ground to complain of breach of contract, or want of good faith on the part of the Postmaster General (quoted in Robertson 1947, p. 69).

Rarely, of course, was there within the British government anything resembling a consensus over plans for a possible nationalization of the telephone system. Some in the Treasury Department regarded nationalization as the lesser of two evils, the other one being fierce and anarchic competition between private firms (Robertson 1947, pp. 55–56). Others saw private—or even municipal—competition as a vital element that would further the successful evolution of the overall enterprise (Perry 1977, 89ff). The constantly shifting positions of influential individuals within the General Post Office were even more complex. Many such influential actors insisted that the telephone had absolutely no future in Britain as a public utility, while at the same time actively urging the government to pursue a policy of nationalization—often simultaneously. A case in point was the Postmaster General Lord Manners. While passionately claiming that the "the telephone could not be utilized on the public wires in any way" (Perry 1992, p. 147), he insisted that "the matter [of private telephony is] not one in which the Post Office could remain passive" (Perry 1977, pp. 73–74).

Yet, despite a constant stream of low-level intra-governmental disputes on the subject, it is unquestionably the case that, by 1880, the small community of British telephone entrepreneurs was becoming increasingly uncomfortable over the prospect of an enforced government takeover. Their fears stemmed from the Telegraph Act of 1869, the nation's primary telecommunications legislation, which was itself a protégé of a series of earlier regulatory settings in the area of telecommunications. These included the Telegraphy Acts of 1868, 1863, and 1854, as well as a number of statutes dating back to 1837, 1710 and 1657—the earliest of which facilitated the state-control of postal services. The language of the Telegraph Act of 1869 left little room for speculation: Section 4, for example, entrusted the Postmaster General with the sole right to operate "any apparatus for transmitting messages or other communications by means of electric signals" (quoted in Robertson 1947, p. 23). Rather predictably, therefore, in 1880 Her Majesty's Government successfully sued the National Telephone Company for privately operating a section of what government lawyers referred to as "the state's electric communications network', meaning the telephone network (Robertson 1947, 21ff; Canes 1966, p. 9). Following a decision by the High Court, the National Telephone Company, as well as a number of smaller, privately or municipally controlled companies, were subjected to a set of strict regulatory guidelines (discussed below). Furthermore, in addition to income taxes, a royalty consisting of 10% of the company's gross annual income was required as payment to the General Post Office (Robertson 1947, pp. 26, 45; Perry 1977, p. 82). Finally, operational licenses became due [to expire] at the end of [31] years, i.e. on 31 January 1912; the Post Office reserved the right to purchase the system at the end of the 10th, 17th, or 24th year; what would happen if, in 1912, the Post Office refused to renew the license was not mentioned (Robertson 1947, p. 26).

The National Telephone Company held out until 1912, when it was finally purchased by the General Post Office (Perry 1977, p. 90). Additionally, following nationalization, only 13 of the 60 municipalities that expressed an interest in developing their own, locally controlled telephone systems, were granted licenses by the state. Of the six that ultimately proceeded to use them (Glasgow, Brighton, Hull, Swansea, Portsmouth, and Tunbridge Wells), only the Hull Telephone Company managed to survive in the long run. In retrospect, says the historian Charles Perry, it seems that "the movement for municipal exchange systems was a smokescreen that only delayed the inevitable creation of a single system under [General] Post Office management" (Perry 1977, p. 89).

3.3   Challenging the Motives Behind Nationalization

On the surface, the British government's takeover of the telephone system constitutes one of the most paradoxical events in the history of the country's economic policy. For Britain, the late nineteenth and early twentieth centuries marked an era permeated by an overwhelming support among governing elites for open markets that were free from substantial state intervention. It was a time when virtually every single pillar of Britain's industrial might, namely coal, bronze, steel and shipbuilding, were revered globally as the products and emblems of free capitalist enterprise. Remarkably, it was within that political and economic context that, first the telegraph in 1869, and then the telephone in 1912, were completely taken over by the state. They were the first nationalization projects in British history; and, as economic historians inform us, so unprecedented were they that they actually prompted the first use of the term *nationalization* in the lexicon of British economic policy (Tivey 1966, p. 65). Even more remarkably, both projects were conceived, designed and implemented, not by the Labour Party—which didn't even exist until 1900—but a series of Conservative and Liberal (meaning centrist) Party administrations (Barry 1965, p. 177). Remarkably, the scholarly literature on the subject is mostly silent. A notable exception is Charles Perry, who presents us with the insightful claim that the roots of this silence are to be found in "the tendency of [...] historians to link nationalization with Labour Party history and, therefore, to dismiss those cases of government expansion which are not part of that tradition" (Perry 1992, p. 86). Indeed, none of the works on British nationalization that are considered seminal by economists acknowledge the telegraphy and telephony nationalization projects of 1869 and 1912 (see Blank 1973; Brooke 1991; Chick 1991; Middlemas 1986; Rollins 1992).

The small number of historians who have examined British governmental attitudes toward telecommunications have, for the most part, been all too willing to distinguish two primary motives behind the decision to nationalize. On the one side, the argument goes, the government aimed to provide an efficient, standardized telephone system under a unity of control, thus accommodating the communication needs of the business sector, as well as of the general public (Pierce 1977, p. 192; Perry 1977, pp. 83–84). Additionally, it is often argued that government and General Post Office executives interpreted the rise of the telephone as a threat to the state's telegraphy service and decided therefore to nationalize the telephone network so as to preemptively recover any lost revenue (Robertson 1947, pp. 22, 45, 87; Perry 1977, p. 82).

The first supposed motive behind nationalization—namely that the government nationalized telephony in order to make it more efficient—appears to be, quite simply, a fallacy. As later sections of this chapter will demonstrate, the British government's takeover of telephony was the primary cause of the *underdevelopment* of telephonic communications, as well as of the systematic discrimination and exclusion of non-business users, which lasted well into the late 1960s.

The second motive—namely that the government nationalized the telephone network in order to protect its telegraphy rights and revenue—corresponds to a greater degree with reality. It is difficult, and indeed unwarranted, to overlook the strong interest by the British government and the General Post Office in safeguarding the dominant status of telegraphy as the major communicative artery of the nation in those days. By 1876, the telegraph network had been in use in Britain for more than four decades. At that time, there were already over 5000 telegraph offices throughout the country and the number of dispatched telegrams exceeded 50 million per annum. Equipped with over 200 long-distance submarine cables, which linked London to the furthest corners of the Empire, Britain undisputedly possessed the finest and most efficient communication system in the world (Aronson 1977, p. 16; Perry 1977, p. 75). Government Ministers and General Post Office executives regularly proclaimed their determination to prevent an experimental voice transmission system, promoted by Bell, a generally unknown Scottish inventor, from displacing the already time-tested telegraph. The latter, they reminded, was the most profitable, most extensive and most used electric communication network on the planet. On one such instance, on 22 March 1893, the Postmaster General, Sir James Fergusson, warned that there was a

> real danger of the valuable public property, the telegraph system, being injured by the extension of this telephone system [. . .]. Wherever the telephone system has been principally developed, there the growth of the telegraph revenue has been checked [...]. If the telephone companies were in communication with all the large towns and sent messages all over the country, undoubtedly the system would to a large extent supersede the telegraphs and consequently largely diminish the telegraph revenue. Therefore, it is [...] essential [...] that the government should have possession of the trunk wires (quoted in Canes 1966, p. 10; see also Robertson 1947, p. 45).

In a memorandum to the Treasury Department that was penned a few years earlier, his predecessor, Postmaster General Lord Manners, had expressed similar opinions on the issue of the coexistence of state-owned and private telephone companies:

I think you will see that the matter [of private telephony is] not one in which the [General] Post Office could remain passive, for if renters of private wires, on the expiration of their agreements, had applied to have their wires fitted with telephones, and their application had been refused, they would, I need scarcely say, have employed contractors to erect fresh wires, and the [General] Post Office would have had old wires thrown on its hands (quoted in Perry 1977, pp. 73–74).

However, argues the economist Michael Canes, concerns that the competition for customers between telephone and telegraph networks might result to a reduction of telegraph revenue for the state did not necessarily have to lead to nationalization (Canes 1966, p. 10). From an economic standpoint, there were many alternative solutions to the government's concerns, including restrictions on telephone investment, the functional limitation of private telephone enterprises to local-distance communication, or even the imposition of high government taxation on the gross income revenue of private telephone companies. Nationalization was but one of many possible measures that could have adequately ensured the survival of telegraphy. Canes' reasoned argument is further-supported by data that demonstrate telephony's poor financial performance under state control: by 1915, the initial surplus of £303,000 that had been recorded in 1913 had been extinguished and replaced with a £100,000 deficit (Raine 1920, p. 95). Indeed, positive financial indicators never reappeared during the next 70 years. It is fair to state here that the absence of parliamentary oversight is often blamed for these poor economic indicators. In 1932, for instance, the Conservative Member of Parliament Roundell C. Palmer protested that the British

Parliament was unable to see what costs and profits accrued separately to the telephone or telegraph postal services. Bureaucratic procedure dominated the organization and ministers overseeing the department came and went with unusual rapidity (Wolmer 1932, p. 14).

But despite such reasonable protestations, there is a lack of logical coherence in the argument that telephony was brought under state control in an age of free enterprise in order to safeguard the survival of another state-owned telecommunications network, namely that of telegraphy. If it were indeed the case that the primary motive behind the nationalization of telephony was to protect the state's telegraphy revenue, then how are we to explain the earlier nationalization of telegraphy? And, if the motive for that was the protection of the state's postal revenue, then how are we to explain the state's postal monopoly, which occurred in the 1600s? Surely, a solution that would have been much more compatible with the spirit of the times would have been to facilitate the privatization of all telecommunications, as well as the provision of a legal framework that would promote the development of the sector through free-market competition. That was manifestly the case with all large-scale industries of the day. The fact that this did not occur cannot be effortlessly dismissed as an instance of economic inconsistency by a series of—primarily Conservative— governing administrations of what was at the time the world's dominant capitalist superpower.

A desire by elements of the British state to safeguard the interests of the public, as well as of business users of the new technology, is also favored by some historians an additional motive behind the nationalization of telephony. Privately operated telephonic communications were expensive and inefficient. Thus, the argument goes, by introducing a much-needed unity of control over telephony, the government hoped to promote its expansion (Perry 1977, pp. 83–4). According to orthodox economic theory, this argument would fall under the category of social regulation of private enterprises; the latter denotes the type of governmental regulatory intervention that primarily aims to promote public interests, which the profit-motivated market is often unwilling or unable to meet voluntarily (Hills 1986, p. 29).

This argument, however, appears to be historically unfounded. To begin with, commentators agree that it was regulatory strangling, imposed by the British government on private telephony following the 1880 decision of the High Court (discussed earlier), that caused the technical and administrative inefficiencies of the telephony network prior to the nationalization of 1912. In London, for instance, no telephone company was allowed to operate beyond a radius of five miles from a central point. At the same time, companies were forbidden from constructing trunk lines, erecting poles on, or passing wires over, private or public property, a restriction that made the practical operation of a telephone system virtually impossible. Hills (1986) has stressed that it was this very

> geographical limitation of these local networks through structural regulation, coupled with recurring threats of nationalization [that] stimulated local monopoly exploitation, which in turn led to public pressure for nationalization (Hills 1986, p. 78).

Reportedly, the situation became so desperate that, on 7 June 1895, the principal technical journal of the country, *The London Electrician*, issued a dramatic editorial that called for the total abolition of the entire telephone system (Robertson 1947, p. 52). It appears, however, that the striking popular, parliamentary and administrative consensus that allowed for the nationalization of telephony had been generated by frustration over the inefficiencies of private telephony, which the government itself had created. It is worth noting that the inefficiencies of the telephone system multiplied following its takeover by the government. An examination of the telephone's development in Britain following nationalization points to the fact that the decision to nationalize was

> at once short sighted, immature and legalistic. [What is more, t]hat it was established so early in the history of the telephone development in this country was one of the major causes of slow progress over thirty years (Robertson 1947, p. 21).

From the very beginning, governmental administration of the telephone network was so tragically inefficient that the notoriety of the service reached unprecedented levels. Writing just 8 years after telephony's nationalization, the British political commentator Gerald Raine remarked that

> delays and mistakes drive the user [of telephony] to exasperation and sometimes to despair. Its inefficiency is a severe handicap to the business of the country [. . .]. The outstanding features of telephone control by the state have been the vast increase in the cost of running and also the cost to the user, coupled with a steady deterioration in the quality of the service (Raine 1920, p. 95).

Political power-struggles in and out of Parliament, as well as the bureaucratic use of red tape by General Post Office officials, resulted in circumstances that were often farcical. On one illustrative instance, the General Post Office refused to purchase telephone instruments from private manufacturers, while at the same time refusing to manufacture them itself (Raine 1920, p. 49). Ultimately, the technical features of the telephone network were, in the words of British political scientist Jill Hills, "consistently undermined" by the Treasury's "lack of investment, [which] precluded the necessary technological upgrading, while lack of marketing kept usage rates low" (Hills 1986, p. 78). The frustration of users is detectable in written records of the time. The aforementioned Gerald Raine exclaimed in 1920 that "[t]hese telephones are so essential to the community and the loss and annoyance that result from this inefficiency are so great, that we may anticipate a resolute movement in favor of their denationalization" (Raine 1920, p. 96).

Long before the General Post Office took over the telephone system, its pioneers in Britain saw it as a technology designed mainly for corporate use (Perry 1977, p. 83). Consequently, from the 1880s to the 1930s, business interests were exceptionally influential in determining telephone pricing policy. It was only in 1915 when the General Post Office hesitantly began to consider the gradual abolition of the flat-rate pricing system and the adoption of reduced telephone charges for the public, at the expense of business users—a policy which in Europe and North America had been brought into force almost two decades earlier (Perry 1977, 91ff; de Sola Pool et al. 1977, p. 131). Yet, even that development did not challenge the elitist culture of telephony administrators. Successive generations of skeptical and uninspired General Post Office officials stubbornly refused to view the telephone as anything other than a luxury that was to serve the needs of business and the aristocracy. According to the communications scholar Carolyn Marvin, the class-based exclusiveness of the telephone service constituted a "political priority" (Marvin 1988, p. 101) for the British government. These elitist sentiments were openly expressed at an early stage by no other than the very Postmaster General, Arnold Morley, who, in a statement before the House of Commons in 1895, argued passionately that "the telephone could not, and never would be an advantage which could be enjoyed by the large mass of the people" [quoted in Marvin (1988)]. Well into the twentieth century, therefore, the telephone remained far beyond the reach of the vast majority of the British population. An annual fee of £17 was required for unlimited telephone use in 1920, when a family could hire and maintain a servant on a full-time basis for £20 per year (Robertson 1947, p. 80, 188; Perry 1977, pp. 78, 83). Inevitably, in December of that year, less than 2% of the total population of Britain were reported to be telephone subscribers (Perry 1977, p. 82).

Combined with the systematic underdevelopment of rural telephony, that underwhelming situation inevitably resulted in the retardation of the British telephone system in comparison to systems in continental Europe, the United States and Japan. In 1927, a local telephone call in Britain was twice as expensive as in the US and three times as expensive as in Norway or Sweden. By 1915 approximately ten million telephones were operating in the United States. That number was only reached in Britain more than half a century later, in 1965. International comparison is, of course, not the only method of confirming the stagnant development of the telephone in Britain. It can also be demonstrated by comparing the extremely low dissemination of telephones against a number of other household appliances, such as refrigerators, washing machines and television sets (Marvin 1988, p. 64; Canes 1968, p. 18; Robertson 1947, p. 34; Aronson 1977, p. 23; Perry 1977, p. 80; Hills 1986, pp. 81, 84; Wolmer 1932, pp. 48–149).

3.4   The Socio-Political Context of the Nationalization of Telephony

Undoubtedly, the fear that telephony might eliminate the lucrative revenue that was generated by telegraphy was very real among some influential British government administrators. Additionally, it cannot be disputed that some British officials believed that the technical and operational consistency of the telephone system would improve under unified state control. It is evident, however, that the policies of the government, both prior to and following nationalization, do not attest to these concerns. It could well be, then, that the primary concerns that informed the British government's telephony policy are to be found elsewhere.

It is impossible to carry out an accurate study of the issue in question without examining, not only the economic and administrative, but also—indeed primarily— the political and ideological context of the nationalization of telephony in Britain. It is likely that some of the more significant motives behind the government's decision to nationalize telephony lie precisely in that context. More specifically, one needs to explore the extent to which ideological factors influenced the takeover of telephony.

The second half of the nineteenth century was a period of heightened class conflict, as Britain was waking to a number of serious social side-effects of its ambitious industrial experiment. The emergence of a new, flamboyant, capitalist elite, alongside the persisting economic degeneration of millions of toilers who were cramming into the working-class ghettos of Britain's industrial centers, were but two elements that were rapidly causing a largely unprecedented political militancy among the country's working and lower-middle classes. The nationwide Chartist movement of the mid-nineteenth century gave rise to a number of reformist tendencies among the population, notably the campaign for women's suffrage, which culminated at the turn of the century. Many of these movements had working-class origins but gained considerable middle-class or even upper-class support, as was the case with Christian socialism, which rose out of the non-conformist Christian reform movement of the late nineteenth century.

But the largest and most powerful reformist movement in British society during the time of the telephone's emergence was the trade union movement that led to the establishment of the Labour Party in 1900. By 1912, the year of the nationalization of the telephone system, organized labor had established itself as a major mobilizing force and a leading actor in the nationwide political negotiation on a number of critical issues (Hinton 1983, p. 1). Alongside the organized labor movement, new norms of even more ambitious organizations had emerged. Militant formations, such as the syndicalist-oriented Land and Labour League (est. 1869), H.M. Hyndman's Social Democratic Federation (est. 1880) and, later, the immensely popular socialist Fabian Society (est. 1884), began to draw together previously disparate radical segments of the British and Irish working and middle classes. Soon, the list of radical groupings and formations of all colors and sizes began to grow by the month. Some of the larger among them, such as the Socialist League (est. 1884), the Scottish Labour Party (est. 1888) and the Independent Labour Party (est. 1893) sponsored numerous radical publications, society clubs and propaganda outlets throughout Britain.

Political militancy ran parallel to a resurgence of nationalism. Since 1858, the first semi-secret Irish republican organization, the Fenian Brotherhood, and its thousands of members in America, had declared war against Britain. Other militant Irish republican organizations, including *Clan na Gael*, emerged at the turn of the century in the United States, Australia and Canada. These groups launched a coordinated campaign of sabotage and subversion against the British state and its interests in Ireland and the world (Porter 1987, p. 26). The colonies were another fertile ground facilitating the growth of anti-British sentiment among native populations. By 1910, nationalist and other radical republican political networks were operating in Southeast Asia, the Middle East and sub-Saharan Africa, promoting the violent overthrow of the British colonial forces. These networks found fertile ground in the more radicalized sectors of the British labor movement, which saw themselves as having little in common with the governing elites of the Empire.

Perhaps the most notable feature of rising working-class militancy in Britain was the striking growth of trade union membership throughout the nation. The latter grew from less than 500,000 in the mid-1870s to over two million by 1899. At the dawn of the new century, Britain witnessed the rise of the most extensive and organized trade union movement in the world: by 1914, trade union members accounted for more than a quarter of the total workforce—upwards of four million workers—in virtually every British industry (Hinton 1983, pp. 24, 84). Total union membership was growing, on average, by one million per year (Wigham 1976, p. 33). Consequently, within 5 years, total union membership had soared to eight million (Morgan 1987, p. 14).

Interestingly, the primary reason behind this unprecedented upsurge in union membership was not the deteriorating conditions in the workplace. In fact, working conditions had improved remarkably since the 1850s (Hunt 1981, pp. 76–98). Rather, it was the increasing administrative capacity of trade union organizations to manage, monitor and steer large numbers of members throughout the nation. In his seminal study of British working class movements, the economic historian Edward Hunt takes after the labor historians Sidney and Beatrice Webb in describing late nineteenth and early twentieth century unions as "new model unions', distinguished from their predecessors by their advances in union organization, communications efficiency and technical skill (Hunt 1981, p. 250; Kirk 1994, p. 76). It was this very organizational efficiency of unions, facilitated by telegraphy, that led to the gradual formation of networks of coordination between geographically disparate labor organizations. Already by 1860, networked trades councils had appeared in every sizeable town and city in the British Isles (Hunt 1981, p. 264), while some unions, such as the Society of Engineers, had set up networks of paid correspondents in virtually every one of the nation's industrial centers (Wigham 1976, p. 2). Other unions had managed to extend their ties to international labor bodies: Karl Marx's London-based International Working Men's Union used telegraphy to coordinate and hold annual congresses in six different European cities (London, Geneva, Lasanne, Brussels, Berne and The Hague) between 1865 and 1872 (Beer 1920, p. 217).

Telecommunications networking gradually encouraged the amalgamation movement, which decisively altered the history of British labor unionism: in 1908 the Miners' Federation of Great Britain incorporated the last independent miners' unions of northeastern England to become the first national union body. The National Transport Workers' Federation (Renamed Transport and General Workers' Union in 1920) followed in 1910, as did the National Union of Railwaymen (1912). Finally, there came in 1914 the Triple Alliance of miners, railway and transport workers, which "provided for the co-ordination of strike action between its constituent unions [and] was widely seen as a portent of a revolutionary general strike movement" (Hinton 1983, pp. 91–92; see also Jeffery and Hennessy 1983, pp. 1–2). At the end of the first decade of the twentieth century, Trade Unions Council annual meetings used telecommunications to involve the almost daily solicitation and coordination of the voting preferences of more than 1.5 million union members located throughout the British Isles (Beer 1920, p. 316). Such long-distance coordination facilitated the systematization of Parliament lobbying, national walkouts, and collective bargaining, which, in the words of labor historian Raymond Postgate, gave unions "an amount of power which would surprise a modern trade unionist" (quoted in Hunt 1981, p. 262).

This newfound power was unleashed during large, carefully coordinated and national, or even general, strikes—a form of organized action that "emerged as labour sought to override the constraints of locality through trade union organization" (Charlesworth et al. 1996, p. xiii). The national lock-outs of 1852 and 1897–1898 by the Amalgamated Society of Engineers; the 9 h movement of 1871, the field revolts of 1872–1874, and the May Days of 1890, 1891 and 1892, which involved the coordinated participation of more than three million demonstrators from Aberdeen to Plymouth, brought back memories of the Chartists and the General Strike of 1842 (Charlesworth et al. 1996, pp. 112–115). But it was the railway strike of 1911 that truly shook Britain: it was the first-ever railway strike on a national scale and the country was brought to a complete halt. The government had to react rapidly, placing every British soldier in the country at the service of railway company managers. The strike was not over until troops were stationed on the streets of most large towns and cities and martial law was declared over most of the nation (Hinton 1983, 86ff). According to the labor historian James Hinton,

[t]his was no ordinary strike movement. The spontaneity and breadth of the strikes posed unprecedented problems [for the government]. And the longer settlements were delayed the more the strikes took on the character of a general social war [. . .]. Troops were called out in several centers, and Salford was subjected to virtual military occupation. In Liverpool, with a gun boat standing by on the Mersey, two strikers were shot dead by the Army following three days of guerrilla warfare in the streets around the city center (Hinton 1983, p. 86).

The government was also alarmed by another new phenomenon, namely that of sympathy strikes, in which networked union groups around the country would walk out in support of a local labor dispute, even if the latter concerned a completely unrelated part of the industry hundreds of miles away. Sir George Askwith, a senior civil servant at the Department of Labour, wrote at the time that "[n]o community could exist if resort to the sympathetic strike became a general policy of trade unionism" (quoted in Wigham 1976, p. 29).

A few months later came the most dramatic labor dispute on a national scale until the 1926 General Strike, described by observers at the time as "the biggest stoppage the world had yet seen" (J. Lovell quoted in Kirk 1994, p. 107): less than 4 years after its national amalgamation, the Miners' Federation of Great Britain called out a national strike on 1 March 1912. It was the first national miners' strike ever and caught the nation totally unprepared. Within days, 600 factories had closed down due to lack of energy, while all rail services were stopped and one million workers had been laid off. Coal prices soon reached famine levels (Hinton 1983, p. 85). Throughout the strike, the British press, as well as the public and many politicians, were under the impression that a revolution, or total breakdown of society, was imminent. Sir Edward Gray, a Member of Parliament, characteristically wrote to a friend:

[t]his coal strike is the beginning of a revolution. Power has passed from the King to the nobles, from the nobles to the middle classes and through them to the House of Commons and now it is passing from the House of Commons to the trade unions [. . .]. The unions may of course, like blind Samson with his arms round the pillars, pull down the house on themselves and everyone else, if they push things too far; or if the owners are unyielding, there will be civil war [. . .] (quoted in Wigham 1976, p. 28).

The above quote is indeed typical of popular sentiment at the time. It signifies the realization in governing circles that union groups had amassed an advanced degree of organizational skill, considerable networking power and large numbers of members who were now "capable of national strike action that could bring the whole economy to a halt" (Hinton 1983, p. 93). By means of that realization, the British state—represented by the Department of Labour and the Cabinet, which until then had adopted a policy of non-involvement in industrial disputes—began to proceed to what British government historians Keith Jeffery and Peter Hennesy describe as "the creation of an official strikebreaking machine" (Jeffery and Hennessy 1983, 3ff). This security machine was exposed in 1919—following yet another paralyzing railway strike—in a report submitted to the War Office by the General Headquarters Great Britain, which had been established shortly before the Great War with the aim of conducting military operations at home (Jeffery and Hennessy 1983, 22ff). The largest part of the report concerns the communications between troops and other emergency services and makes particular mention of telephones as valuable instruments for monitoring and combating strikers. There is also a request that, in their efforts to control union strike activity, officers "should have the right to use public call telephones without payment', since the latter now belonged to, and were used for the interest of, the government (Jeffery and Hennessy 1983, p. 22).

The effective utilization of the telegraph and the telephone were the technical means that facilitated the increasing nationwide coordination of union activity. Instantaneous communications would have indeed been the only means by which synchronized, collective action involving the simultaneous involvement of millions of union members, organized in local associations throughout the British Isles, would have been feasible or indeed possible. The dramatic potential for "reorganization of the social geography" (Marvin 1988, p. 66) that was offered by the new, instant communications systems was used to augment the coherence, timing and efficiency of organized campaigns by labor groups. At the same time, the government became increasingly interested in employing the instant communications network for the purpose of coordinating its response to such all-encompassing subversive activities—as illustrated by the aforementioned 1919 report to the War Office, and as will be explained in the next chapter.

Remarkably, the relevant research literature shows no consideration of the nationalization of telephony in the context of the rapid radicalization of British society during the late nineteenth and early twentieth centuries, or indeed in the context of the increasing governmental intervention in industrial disputes. For the most part, it views the telephony's nationalization as a fundamentally financial decision—an extension of the nationalization of telegraphy. The telegraphy nationalization debate, which took place in the late 1850s and early 1860s (Perry 1992, p. 91), was indeed echoed by the telephony nationalization debate, which followed a few years later. Public frustration mounted over the high dispatch fees that were charged by privately owned telegraph companies, which limited the use of the system to a small, wealthy elite (Kieve 1973, p. 126). There was a class of civil servants who viewed nationalization as the state's means of providing a social service, spreading telegraphy to rural areas, and uniting the nation through the immediacy of telegraphic contact (Perry 1992, pp. 89, 111, 119). Finally, there existed amongst legislators the belief that telegraphy was too significant an asset to be abandoned at the hands of profit-oriented corporations. In the words of Frank Ives Scudamore, Accountant General of the General Post Office,

the transmission of news [...] throughout the Kingdom should be regarded as a matter of national importance and the charge for such transmission should include no greater margin of profit than to suffice to make the service fairly self-supporting (quoted in Perry 1992, p. 105; see also Kieve 1973, p. 146).

Yet, if we were to base our analysis of the nationalization of telephony on the precedent of telegraphy, we would have to consider a number of very real non-financial factors that influenced the decision to nationalize cable communications. The research literature reveals a debate running in parallel with the more widely-distributed discussions on financial inclusion and social service provision, namely the debate on the political significance of the state's control of the communications networks. It is often overlooked that, before the first Telegraph Act of 1854 was even debated in Parliament, telegraphy had already been nationalized in a number of British colonies, including India. It was there that the British government was able to explore the military command, control and communication uses of the new technology. James Dalhousie, the London-based Governor General of the East India company, was instrumental in the nationalization of telegraphy in India. He viewed the technology, not as a business venture, but as an instrument of British military might—a device that had contributed to "the early realization of a vast magnitude of increased political influence in the East" (Dalhousie quoted in Headrick 1991, p. 52). In 1854, soon after the proposal for the nationalization of telegraphy in India was enacted by Parliament, Dalhousie wrote the following to a friend:

> [t]he post takes ten days between [Calcutta and Bombay]. Thus in less than one day the government made communications which, before the telegraph was, would have occupied a whole *month*—what a political reinforcement this is! (quoted in Headrick 1991, p. 52).

In 1857, the Empire's military commanders had the opportunity to decisively test the telegraph in action during the large Indian army mutiny, which engulfed most of northern India and was the strongest challenge to the British Empire for over a century. The outnumbered and outgunned British forces used the new communications instrument to surround and capture the mutineers. Following the crush of the rebellion, John Lawrence, Commissioner for the Punjab, was quoted as having declared that "[t]he telegraph saved India" (Headrick 1991, p. 52). The American historian of technology Daniel Headrick informs us that, following the subversion of the Indian rebellion, the telegraph spread in all regions of the British Empire largely due to "the military and political needs of the British rulers" (Headrick 1991). British Africa was among those regions. In 1877, Thomas Watson, president of the Cape Town Chamber of Commerce, wrote to the Cape's governor arguing in favor of the establishment of telegraphic communications with London:

> [t]he construction of a line of telegraph through the center of this great country would not only put us in immediate communication with the mother country, but at the same time open up a vast field for commercial enterprise. The maintenance of a series of stations along the route would do more to abolish the slave trade than [...] a fleet of cruisers on the African coast. Mission stations would be protected, savage tribes civilized and in a few years a complete revolution would be effected (Headrick 1991, pp. 62–63).

It was also thought that requirements relating to the coordination and defense of the Empire at large, at that time more geographically stretched than ever before, would be met through the use of electric telecommunication systems. According to Irish historian and philosopher W.E.H. Lecky, the use of the telegraph had "at least modified the Irish difficulty by bringing Dublin within a few minutes' communication of London" (quoted in Marvin 1988, p. 98). Celebrating the initiation of the operation of an 8100-mile telegraphic circuit connection from New York to London in 1880, the British Special Commissioner to Canada, Henry Norman, characteristically stated: "[i]s not the click of this key, heard in two hemispheres, more eloquent than all the arguments of empire ever penned?" (quoted in Marvin 1988).

As the subsequent dismantlement of the British Empire shows, such hopes did not materialize. The establishment of electric interconnectivity between the Empire's main strategic centers did not prevent its unceremonious collapse. Such hopes were indeed instrumental, however, during the national debate on the nationalization of telegraphy. Marvin agrees that it was not long after the appearance of electric communication that the British state recognized its significance for political control. It was often presented as "a superior means for controlling masses, criminals, primitives, servants, and whatever other underclasses might need restraint', she writes (Marvin 1988, p. 100). According to *The London Electrician*, riots and other "unruly meetings" that took place every winter in West London could be countered by the empowerment of electric communication, demonstrated by its ability to collect and coordinate a large police force "at any desired point, within a very short time" (7 July 1888, quoted in Marvin 1988, p. 99). The Irish historian W.E.H. Lecky wrote in 1896 that electric communications had "greatly strengthened the central government in repressing insurrections, protecting property, and punishing crime" (quoted in Marvin 1988).

In 1861, legislators J.L. Ricardo and W. Burchell produced a report in which they accused the private telegraphy companies of being "individual and irresponsible" (quoted in Perry 1992, p. 90), while asserting that publicly-elected statesmen should be in charge of enhancing the system's military and diplomatic usefulness (Perry 1992). Others contributing to the debate directed attention to the usefulness of the electric telegraph as a means of communicating and acquiring intelligence (Kieve 1973, p. 124). It was in fact precisely for that reason that many concerned actors and observers positioned themselves against the nationalization of the telegraph. The danger of government espionage against its citizens "as practiced in Europe" was one of the private telegraphy industry's primary arguments against nationalization (EITC 1868, pp. 12–13). The national and local press repeatedly raised the issue (Kieve 1973, p. 145) and a Parliamentary committee was set up to investigate "what securities should be taken for insuring the secrecy of messages transmitted through the [General] Post Office" (quoted in Kieve 1973, p. 148). A member of the committee, Member of Parliament George Leeman, raised the issue of state agencies potentially intercepting telegraphic communications between subscribers by reminding Parliament of the notorious scandal of the interception of Giuseppe Mazzini's mail correspondence by the police, a few years earlier (Perry 1992, p. 108). The Mazzini case was the first ever public controversy over state-sponsored communications interception in Britain. It occurred in 1844, when it surfaced in the English press that the Home Secretary had issued a warrant for the interception of the correspondence of the Italian revolutionary Giuseppe Mazzini, who resided in London (Baxter 1990, p. 161).

The parallel between postal correspondence and telegraphic communication is indeed quite appropriate. The social historian Jeffrey Kieve has noted that, through-out the debate on telegraphy's nationalization, "[i]t was argued that many of the reasons for the Post Office being made a government department seemed equally applicable to the telegraph" (Kieve 1973, p. 124). Kieve's observation provides a crucial link in outlining a security and policing parameter in the British debate over the nationalization of communications. Scholars are in agreement that the 1657 Ordinance Act, which was enacted during the Cromwellian period and which established the British government's postal monopoly, was primarily a security measure aiming to enhance the republican government's control over the content of postal correspondence. The Ordinance Act itself describes the monopoly as "the best means to discover and prevent any dangerous and wicked designs against the Commonwealth" (Bunyan 1976, p. 197; Baxter 1990, pp. 160–161; Fitzgerald and Leopold 1987, p. 36). According to Sir Edward Murray, former General Secretary to the General Post Office,

> [t]he original object of the state monopoly was not so much to extinguish competition as to give the government of the time access to the correspondence of suspected persons, and particularly to letters passing between England and foreign parts (Murray 1927, p. 3).

This significant political element in favor of governmental control of communications, was clearly present in the postal and telegraphy monopoly debate, as it was present in the discussions around the nationalization of telephony. In his PhD thesis entitled *Ideology and the Telephone*, Jeremy Stein summarizes the argument by pointing out that the years between 1905 and 1911 witnessed "a great awakening to the value of the telephone service', not only for business efficiency, but also for the ability of government to make use of telephony in order to "keep page" with a rapidly changing political landscape (Stein 1996, p. 85). By then it was clear that the telephone's "capacity to bind space and to break down social as well as spatial barriers" (Kern 1983, 208ff) rendered it "an essentially democratic technology" that could "either legitimate the status quo or [...] effect change" (Stein 1996, p. 123). The British government's efforts to shape the direction of telephony's evolution was therefore part of a wider "attempt [...] in a period of great social unrest, to retain control of a sensitive means of communication" (Stein 1996, p. 17), he points out. Stein perceptively concludes that "the Telephone Question', as *The Times* titled it in 1897, was "a highly political process" (Stein 1996, pp. 137, 123). As will be shown in the next chapter, this process was partly informed by the perceived need by law enforcement and security services to monitor the use of this new technology by political radicals, criminals, and foreign spies. It was also informed by military perceptions of telephony as a technology that could revolutionize interstate warfare (Herrmann 1996, pp. 73–74). Ultimately, the near-complete absence of ideological and political factors from the scholarship concerning the nationalization of the telephone network in Britain is not proof of their non-existence; rather it is a manifestation of the way in which economic-oriented deterministic accounts of the nationalization process have been favored by scholars at the expense of a more holistic approach that includes the incorporation of the role of political and ideological elements in the process.

There are, of course, many more crucial issues that remain unresolved: how does, for instance, the nationalization of telephony relate to the expansion of the social regulatory role of the British state during the first decades of the twentieth century, and especially the coordination of the extended suffrage after 1906 (Pugh 1978, pp. 29–44)? How are we to explain the nationalization of telephony in light of the gradual creation of a war economy in Britain, immediately prior to the outbreak of the Great War—a development which also signaled the government's takeover of wireless broadcasting under the Defense of the Realm Act (Headrick 1991, p. 145; Briggs 1961, p. 20)? Equally importantly, what was the precise position of the British military—a particularly vocal actor in all aspects of communications regulation (Hughes 1981, 35ff) —in the overall debate on nationalization? Finally, how did the moral question behind the explosion of gambling and prostitution, two industries which made frequent use of the telegraph and telephone networks to organize their business and attract customers, influence the debate on telephony nationalization? Prostitution and gambling were in fact virtually revolutionized with the adoption of the telephone. By the late 1880s no self-respecting brothel lacked one or more telephone instruments through which customers made their reservations. At the time of the emergence of the telephone, approximately 50,000 prostitutes were operating in the greater London area alone (Thompson 1968, p. 59). The immense impact of the new communications medium on prostitution can also be observed in the emergence of a completely new term: the more traditional words *harlot* or *prostitute*, were replaced by the expression *call girl* (de Sola Pool et al. 1977, p. 136). Gambling, already forbidden by law from advertising in most print publications, switched to the adoption of the telephone as its foremost operating medium (Pertrow 1994, p. 259; see also Munting 1996, p. 27). In fact, it was often the case that such ventures were stationed abroad, beyond the reach of British law enforcement. The use of the telephone network by gamblers and call girls became so extensive that, at the turn of the last century, the very influential National Anti-Gambling League, along with numerous other anti-gambling and anti-prostitution organizations, accused the National Telephone Company of knowingly acting as "commission agents" and "messengers" of illegal and indecent operations stationed in Britain and abroad (de Sola Pool et al. 1977, p. 136; Pertrow 1994, p. 265). This moral frustration generated one of the most tension-ridden single-issue campaigns of the time—and one that can be in many ways compared to the American alcohol prohibition movement, which followed a few decades later (Pertrow 1994, 250ff).

None of the above questions are even acknowledged in the relevant literature. Yet it was within this very social context, at a period of rapid social and political change at all levels, that the nationalization of telephony was implemented. The fear of a potential loss of social control over the organized, underprivileged manual workers; the latter's apparent rejection of bourgeois common sense, social norms and morals; the threat against British global superiority in an increasingly competitive international market; as well as the mounting resentment by Irish and other colonized peoples against British imperial rule, were all partial ingredients in an explosive mixture of threatening socio-political insecurity that dominated British elite sentiment. That deep feeling of insecurity was anything but pacified by the arrival of the telephone. The new voice network proclaimed a radically different mode of social and political negotiation—an interconnectivity that could be used to connect and mobilize critical masses of dissidents who were previously scattered along the four corners of the nation—indeed the Empire. It also represented a dramatic "reorganization of social geography" (Marvin 1988, p. 66) that had the potential, if utilized in certain ways, to augment the coherence and efficiency of political protest, as well as all sorts of indecent venturing.

An examination of the takeover of telephony through the prism of the social context expressed above would have an enlightening effect on our current understanding of the British government's decision to nationalize voice telecommunications. The decision to do so sought to promote and safeguard the interest of the nation, as was indeed often argued at the time (Robertson 1947, p. 45), though not just in economic or administrative, but also in social and political terms. It was therefore not simply technical data and revenue statistics, but also political convictions and ideologies, as well as competing visions of the national interest, that led to the decision to nationalize the telephone network.

3.5 United States: Regulating Deregulation

It has been amply documented that, in the United States, the period immediately after the establishment of the Bell Telephone Company in 1876 was characterized by ruthless competition between inventors, venture capitalists and investors, and marked by the near-complete absence of governmental regulation (Cohen 1992, p. 22). Striking indeed during those years is the frequency with which Bell was called upon to testify in lawsuits filed by individuals claiming to have invented the telephone before him. The long list included, apart from Elisha Gray, the American inventor Thomas Edison. Notably, although several lawsuits reached the United States Supreme Court, the latter upheld Bell's rights in all cases. It is reasonable to assert that the early assumption of the Bell Telephone Company's monopoly control over the American telephone market was a direct result of the inventor's ability to soundly affirm the legal validity of his patents (Cohen 1992, pp. 23, 24).

The relaxed posture of the American government during those early years should not be interpreted as an indication of passivity. It would, of course, be erroneous to attempt to distinguish a monolithic motivation behind the government's stance, especially since the latter manifested itself in a series of rapidly shifting positions and policies throughout the late nineteenth and early twentieth centuries. The following example is illustrative: in 1895 Bell's monopoly patents expired, leading to a decade of fierce competition between rival telephone companies. Being by far the most powerful among those, the American Telephone and Telegraph Company (AT&T) initiated a voracious wave of successive takeovers aiming toward the establishment of a nationwide telecommunications network. It was in 1913, therefore, when the United States government brought forward a federal anti-trust lawsuit against the company. Yet shortly afterwards, while it was engaged in the process of negotiating an out-of-court settlement, the government arrived at the conclusion that the peculiar nature of the telephone enterprise meant that the public interest would be favored by the economic integration of the system. Consequently, as it is often put by historians, an AT&T monopoly was "not discouraged" by government (Canes 1966, p. 12).

The government's decision to allow AT&T to develop in a protective, monopolistic environment permitted the company to assume and retain centralized control of what was to rapidly become a most profitable market. In return, the state reinforced its regulatory role as the *ad hoc* guarantor of the development of an essentially private, capitalistic undertaking (Garnet 1985, p. 3; Temin and Galambos 1988, p. 12). In this manner, the government was in a position to oversee and press, not only for the successful application of the telephone system onto the life of the nation, but also for the development, compatibility and overall coordination of an enterprise that was seen to be an important national asset.

It is a truism, and it is routinely indicated in the relevant scholarly literature, that the very history of telephony in the United States is a study on the structural centrality of state institutions in a deregulated environment (Teske 1990, 17ff). As early as 1907, a series of states began to form public utility commissions. These were regulatory bodies with the responsibility of countering possible price-fixing and other market abuses by monopolistic enterprises, including AT&T. By 1913, public utility commissions were overseeing the AT&T's operations in no fewer than 34 American states, with at least a dozen more considering the formation of similar public regulatory bodies (Garnet 1985, 130ff, p. 191; Teske 1990, p. 2). In 1934, the Interstate Commerce Commission was succeeded by the Federal Communications Commission (FCC) as telephony's primary federal regulatory body. Soon afterwards, AT&T was reaching its target of owning approximately 80 per cent of the country's telephones—a share that it retained until 1982—thus becoming virtually synonymous with the national telephone network (Teske 1990, p. 2). Yet, the acceleration of the company's oligopoly was accompanied by a parallel acceleration of the state's regulatory power, which often directly antagonized AT&T. In fact, scholars often neglect to mention that the United States government actually nationalized the telephone industry for one year in 1918, "under the exigencies of World War I" (Janson and Yoo 2013, p. 983), before returning it to the AT&T monopoly. The contradictory relationship between capitalist competition and monopolization was expressed in a series of successive antitrust lawsuits brought against the company by the government. In the meantime, wishing to keep AT&T on a state of constant alert and caution, the government often encouraged debates centered on nationalization within its circles (Cohen 1992, p. 23), so much so that a contemporary observer remarked the following in reference to the government's antitrust efforts against AT&T:

[w]hether the United States shall continue a democracy, or shall revert to a form of oligarchic control, imposed by corporate states through assumption of state powers, today hangs in the balance (Danelian 1939, p. 380).

After World War II, AT&T continued to function and prosper under the—mostly protective but at times restrictive and penalizing—shadow of the state. Routine contradictory decisions by government, as well as decisive policy splits within its ranks, stemmed in part from the skillful balancing of often competing principles, such as the utilization of telephony in the public interest, political control over a powerful communications medium, market competitiveness, and national security. Gradually, a framework of co-operation between the state and AT&T emerged as the outcome of constant negotiation between them, as well as their will to embrace economic, cultural and political goals that satisfied both. The American political scientist Jeffrey Cohen points out that these goals were

> not always [...] easy or possible, and [...] often led to conflict between the government and AT&T. However, it [was usually] in the interests of both parties to find a mutually acceptable position because of the impact of both actors on each other and their desire for amicable relations (Cohen 1992, p. 20).

3.6 American Telephony and the National Interest

The adoption of mutual targets by AT&T and the state was epitomized in their functional co-operation in pursuit of the national interest, defined here as a set of national and international parameters favoring the successful exercise of military, political and economic equilibrium within a nation-state. The close political relationship between successive American governments and telephone companies is widely known and has been extensively illustrated by historians and other researchers. According to Martin (1991) (p. 46), "strategic political connections" between the Bell Telephone Company and the US government began forming as early as 1877, before the company's incorporation. The "unique relationships" (Temin and Galambos 1988, p. 9) between the company and state, as well as federal, policy centers at all levels created a "peculiar framework" (Aronson 1977, p. 7) of intimate institutional relations, under which managers of AT&T, International Telephone and Telegraph (IT&T) and other telephone companies operated for many decades.

The efficacy of that institutional framework was strengthened, not only by the activities of powerful political allies in Congress (Temin and Galambos 1988, p. 25), nor simply by the armies of political lobbyists operating in Washington DC (Sampson 1973, 183ff), but also by the frequent exchange of technical and administrative staff between government agencies and the telephone companies. During World War II, for instance, the United States Department of Defense was employing considerable numbers of former, as well as future, IT&T employees (Sampson 1973, p. 38). According to Sampson (1973) (p. 183), there were always extraordinary numbers of former Pentagon officers within the ranks of IT&T staff: "in 1968 they included three rear-admirals, two brigadier-generals, 22 colonels and eight captains'. What is more, during the 1970s, the IT&T executive board included John McCone, former Director of the Central Intelligence Agency (CIA) (Sampson 1973, p. 99).

At the end of World War II, Paul Porter, Chairman of the FCC, made the following strategic remark:

the managements of international communications companies are [today] in a position to shape our international communications policy through their ability to negotiate and make arrangements with the representatives of foreign governments (quoted in Sampson 1973, p. 22).

The case of IT&T is particularly illustrative of this. The company, undoubtedly one of the world's largest, with offices in more than 140 countries, was actively involved in the defensive and offensive plans of the United States government during the Cold War (Sampson 1973, p. 56). In the immediate postwar period, prior to the nationalization of IT&T's ventures in the communist republics of Eastern Europe, the CIA had placed agents in the company's offices throughout the Eastern Block, under the full knowledge and approval of Sosthenes Behn, the corporation's founder and president. More agents were placed in IT&T branches in Latin America, including in Bolivia, Paraguay and Argentina (Sampson 1973, p. 41). Scholarly research shows that, on at least two instances during the Cold War, in Hungary and Chile, "IT&T *made* American policy as much as it followed it" (Sampson 1973, p. 269).

After the end of World War II, in anticipation of a subsequent world war fought between capitalist and communist countries, and in an attempt to act in defense of American interests, Behn secretly ordered the corporation's managers in Budapest to slow down and even sabotage production in every possible way. His decision was made "with the understanding of the US General Staff" (Sampson 1973, p. 51). In 1949, Budapest-stationed IT&T Eastern Europe Area Manager Robert Vogeler was arrested by the Hungarian military police on charges of carrying out espionage on behalf of the United States. During his trial, Vogeler explained in his own words that

the co-operation between the [Hungarian branch of IT&T] company and the [United States] military was such that the latter could control the operations of the company (quoted in Sampson 1973, pp. 50, 53).

A little more than two decades later, on 21 and 22 March 1972, *Washington Post* journalist Jack Anderson revealed leaked IT&T documents showing that the corporation had, two years earlier, plotted with the CIA to block the election of Salvador Allende, Chile's leftist president. Fearing a possible nationalization of its ventures in an Allende-governed Chile and in a potentially Marxist-dominated Latin America, the company had tried to sabotage production and help encourage a military coup. In return, said Anderson, the United States government dropped a series of antitrust lawsuits against IT&T (Bertrand Russell Peace Foundation 1972, p. 4, 238; Sampson 1973, p. 232). As can be expected, there was no statement by the CIA regarding the disclosure. However, IT&T executive board member and former CIA Director John McCone admitted that IT&T executive personnel and representatives of the US government had indeed discussed—though not acted upon, he said—plans to undermine Allende's candidacy (Sampson 1973, p. 238).

3.7   Regulatory Discretion as a Platform of Negotiation

In the United States, unlike in Britain, the decisions of successive government administrations not to nationalize telephony appear to have been in harmony with the predominant free-market doctrines of the time. The appearance of telephony was interpreted more as a supplement to telegraphy and less as a threat to it. Additionally, the absence of any systematic revenue-oriented state intervention in telegraphy allowed the government to adopt a discreet role among concerned actors, while encouraging private entrepreneurs to take investment initiative and spread the use of the telephone throughout urban centers.

Yet it would be erroneous to interpret the state's regulatory discretion as lack of interest in the political and social uses of the new invention. The dominant view in American policy-maker circles was that telephony was a natural monopoly, meaning that having more than a single company operating in a specific market would be detrimental to the consumer body due to the financial, social and geopolitical features of the American economy. That view, which generated AT&T's private monopoly, was—among other things—a platform of structural interdependence and policy negotiation. It was through that platform that successive US administrations were able to effectively promote the incorporation of the state's political, diplomatic and security concerns into the broader spectrum of AT&T's market ambitions. Gradually, the very same platform was used by the corporation and its international subsidiary, IT&T, to promote US diplomatic agendas that were in agreement with their own market-based executive decisions.

3.8   Summary: Regulation as a Political Choice

In both the United Kingdom and the United States, the early history of telephony was marked by constant negotiation between concerned actors attempting to define the expediency of telephony, as well as its function as a medium of social negotiation. The debate was characterized by a dramatic array of rapidly changing policies and shifting positions, as well as by the intervention of governmental institutions, technicians, politicians, and business conglomerates. The social and economic priorities of governments in Britain and America have varied tremendously. In Britain, the principles of open market and free competition, upon which the country's economy had been built, were sacrificed in favor of the national interest and—almost certainly—social harmonization. In the United States, the fragile equilibrium of economic competition and state supervision was pursued in an attempt to ensure widespread access to the telephone.

In both cases, however, the role of the state in shaping and defining the features of the new technology has been enormous. It manifested itself through diverse policies: in Britain through the complete and incontestable takeover of the industry by the General Post Office, while in America by rigid and constant regulatory surveillance exercised by the federal government and by individual state administrations. Though distinct, both sets of policies display a common unifying element: they were formed and operated in conjunction with the traditional characteristics and functions of the state as a historical force. Their goals, far from the pursuit of abstract notions of right and wrong, were "deduced from history and enacted into law" (Pierce 1977, p. 193). Ultimately, no nationwide communications network could be implemented without the political approval and close administrative supervision of the executive vanguard of the nation.

It has been claimed that, due to its technical features, no industry renders itself more readily applicable to state management than telephony (Mavor 1916, pp. 2–3). Similar to other public utilities, such as gas, electricity, and water distribution networks, the argument goes, the telephone service is a "natural monopoly" (Pierce 1977, p. 192). That argument dangerously excludes the complexities of monopolistic regulation, for it reduces the latter to a matter of technical necessity. The fundamental nature of the state's involvement in telecommunications is not only technical, but also—and in some cases primarily—social and political (Canes 1966, pp. 12–13). Fundamentally, therefore, state regulation over telephony in Britain and America was dominated "by changing ideology, not changing technology" (Temin and Galambos 1988, pp. 7–8).

Chapter 4
The Interception of Communications in Historical Context

In attempting to elaborate on the particular features of the intricate relationship between telecommunications and security agencies, a vital dichotomy must be made between the *communications of security* and the *security of communications*. The former signifies the uses of telecommunications by security bodies for the functional and administrative enhancement of their organization, while the latter involves the expansion of the security agencies' mission from the social realm to the realm of—postal, then electric, and eventually electronic—telecommunications.

The present chapter is structured to reflect the aforementioned dichotomy with regard to telephony. The first section, therefore, examines the early attitudes of law enforcement and security agencies concerning telephony, and delineates their early development, in an attempt to explore the initial interactions between state security apparatuses and voice telecommunications in the United Kingdom and the United States. The second section engages in a discussion of British and American historical experiences of government-sponsored communications interception, from the time of the emergence of telephony until the end of the Cold War. The focus here is on exploring the phenomenon of information security as an element in the institutional relationship between security agencies and the telecommunications industry. That relationship was significant in facilitating and shaping institutional practices of communications interception for most of the twentieth century.

4.1 Britain: Bobbies on the Line

In the British Isles, the police appeared as the local version of a Continental concept that emerged from the ashes of the French Revolution of 1789. Historians essentially regard it as a two-headed creature, established, on the one hand, to help facilitate the state's security mandates, while, on the other hand, to assist in the implementation of citizen, electoral and welfare rights (Smith 1985, p. 23; Taylor 1997, p. 33). The first organized police force in the British mainland (the Royal Irish Constabulary had

© Springer Nature Switzerland AG 2020
J. Fitsanakis, *Redesigning Wiretapping*, History of Information Security,
https://doi.org/10.1007/978-3-030-39919-1_4

been founded in 1814) appeared in the streets of London in 1829 (Leonard 1964, p. 3; Smith 1985, p. 23). Following plans of the ruling Whigs to introduce a national system of policing (Philips and Storch 1994, p. 80; Taylor 1997, p. 23), law enforcement departments were gradually introduced in other cities, as well as in the countryside. The Criminal Investigation Division (CID, founded in 1842) and the Security Service—or MI5—(founded in 1909), as well as the Special Branch (founded in 1917), were later established for the purpose of criminal and political surveillance (Verrier 1983, p. 26; Fitzgerald and Leopold 1987, pp. 37–38).

Historians tend to differ on the degree to which British security forces throughout the country were keen on the adoption of electric communication. Some indicate that Scotland Yard was telegraphically connected to the various district police stations in London as early as 1849, less than 5 years after the invention of the telegraph (Diffie and Landau 1998, pp. 117–118; Leonard 1938, p. 1). Others stress that, in cases where electric communication and other technological appliances were successfully embraced, their adoption was usually the result of individual visionary chief constables (Whitaker 1964, Chap. 2). Elsewhere it is asserted that, even where "the will to mechanize" was generally present, it was usually hampered by lack of financial resources (Emsley 1996, pp. 75, 150). Finally, some propose that the relevant case studies point to the snail's pace at which technological modifications in the organization of security bureaucracies tend to take place, irrespectively of whether individual chiefs or officers are in support of such changes (Weinberger 1995, p. 37). There is considerable evidence in favor of the latter proposition: from the turn of the century and even after World War I, numerous chief constables and the Home Secretary himself often expressed opinions against the adoption by the police of new mechanical devices, including telephones (Weinberger 1995, p. 37; Emsley 1996, p. 149). Consequently, in a number of large industrial cities, such as Liverpool, police forces resisted the installation and use of telephones until 1908 (Weinberger 1995, 37n). Even as late as 1923, "nearly 200 police stations and about half of the 5000 police houses were not on the telephone at all" (Weinberger 1995).

There is indeed no reason to believe that security agencies were more willing to adopt telephone apparatuses than any other inflexible, bureaucratic institution ever was. Yet, there are numerous reasons to accept that, for all its technophobia and institutional stiffness, British police was much ahead of other government agencies in the application and use of electric communications systems. It is certainly impressive, for instance, that by 1867 all principal and sub-divisional stations in the country were interconnected through the police's private telegraph network (Smith 1985, p. 41). Additionally, it should be noted that the primary argument against the speedy adoption of telephonic communications by British police was not based on a denial of the ability of telephony to facilitate nationwide coordination, but that such an ability was already facilitated by the telegraph network. It was for this very reason that police departments in many smaller towns and cities, which had not previously developed extensive telegraph systems, installed and used telephones almost as soon as the service became available (Leonard 1938, p. 9). Certainly by 1933 a private telephone system connected all divisions, subdivisions and police stations throughout Britain, from the Chief of Police to the last officer on the beat, through a system of call boxes scattered along the beat routes of police officers in every city, town and village in Britain (Leonard 1938, p. 408; Walker 1997, p. 257; Rubinstein 1973, p. 16; Weinberger 1995, p. 36; Emsley 1996, 149ff).

4.2 United States: Telephones and Alcohol

In the United States the police service, as a social concept and institution, was modelled largely on the British system (Caiden 1977, p. 22; Diffie and Landau 1998, p. 9). In populous New England cities, such as Philadelphia, Boston, and New York, police forces were established between 1834 and 1838. Yet officers did not wear uniforms and functioned more like traditional community watchmen, rather than members of an institutionalized security agency. It was during the Civil War that internal political and social unrest led police forces throughout the United States to aspire to a more militaristic organizational model (Caiden 1977, p. 22; Fuld 1909, Chap. 1). Following the Civil War, police departments were established all over the country as semi-autonomous law enforcement bodies (Leonard 1964, p. 18). In the years leading up to the twentieth century, the growth of the anarchist movement, as well as the emergence of the organized labor movement, brought about the appearance of political surveillance agencies in urban and some rural areas (Donner 1980, p. 32). The FBI was established in 1908, while its Intelligence Division began operating in 1919. The latter targeted primarily anarchist, communist and other radical political groups (Kessler 1993, p. 65).

Rather expectedly, the immensity of the American landscape was one of the factors that promoted the relatively swift adoption of telephonic communication by the country's law enforcement agencies. Having already adopted the use of dial telegraphs—which enabled those not knowing the Morse code to send messages over the wires—as early as 1858, and of Wirephoto—an early method for the electrical transmission of images—soon after 1900, American police departments joined some of the earliest groups of telephone users (Marvin 1988, p. 97; Leonard 1938, p. 2; de Sola Pool et al. 1977, p. 138). By 1878, less than a year after the establishment of the first private telephone company, numerous police departments in Boston, Washington, and New York City had been multiply connected to the network. In that same year, the Chicago and Baltimore police departments established a network of 170 telephone boxes scattered throughout their cities for the use of police officers on the beat (Marvin 1988, p. 97; Leonard 1938, p. 9; Diffie and Landau 1998, pp. 117–118; de Sola Pool et al. 1977, p. 138). According to the historian Vivian Leonard, the success of police telephone boxes in Chicago "was so rapid that by 1893 no fewer than 1000 street stations had been installed all over the city" (Leonard 1938, p. 11).

There was considerable resistance within police circles to the adoption and use of telephony in law enforcement operations (Leonard 1938, pp. 6–7, 80). Very often, elements within the state and federal police administrations were prepared to disregard their skepticism only after advantages promised by the proponents of the new technology had been aptly demonstrated (de Sola Pool et al. 1977, pp. 137–138). Undoubtedly, the opportunity for demonstration was offered by the wave of strikes that swept the nation in 1892, including the violent railway workers' strike in the city of Buffalo, New York, where combined police and military operations played a crucial role in neutralizing organized protests. *The Electrical Review* later remarked that the "newly established telephone apparatuses [were] invaluable aids" in recalling law enforcement and military officers from their residences (quoted in Marvin 1988, p. 99).

Assertions that the telephone service was an invaluable tool for social control (Martin 1991, p. 145) and that it gave law enforcement and security agencies a significant advantage over organized crime (de Sola Pool et al. 1977, p. 137), were regularly propounded by police and public officials. Such proclamations came to be backed by the experience of the Prohibition of liquor, as well as of illegal gambling, in the decade following 1913, when Congress enacted the Webb-Kenyon Act. Among other things, the Act forbade the mailing or shipping of liquor into states that banned such shipments. In December of 1917, Congress approved the 18th Amendment to the Constitution, prohibiting the manufacture, sale, transportation, import and export of liquor. It was ratified by the states in January of 1919, and in October of 1919 Congress adopted the Volstead Act, which provided for the nationwide enforcement of the 18th Amendment and defined intoxicating liquors as those containing at least 0.5% alcohol. The 18th Amendment went into effect during the following year. By a remarkable historical coincidence,

> Prohibition coincided with the telephone system's years of growth to a national network and total penetration [. . .]. The bootleggers and the rackets made full use of whatever was available to run their operations (de Sola Pool et al. 1977, p. 28; see also Martin 1991, pp. 44–45).

In the press there often surfaced accusations that criminal motives were behind large portions of telephone usage. The New York Telephone Company was accused in 1907 of earning more than 2 per cent of its annual revenues—approximately $1 million—from telephone calls relating to gambling and other illegal operations by its users (de Sola Pool et al. 1977, p. 137). The telephone lines, therefore, increasingly became subjects of confrontation between the police and organized crime syndicates that operated underground. The relevant scholarship shows that telephone surveillance was the principal method and the most frequently used tool of American law enforcement agencies in their effort to enforce Prohibition measures throughout the nation (Dash et al. 1959, p. 28; Diffie and Landau 1998, p. 28). Indeed, as the telephone apparatus began to mirror the pattern of everyday life in all its expressions, law enforcement agencies found it difficult to abstain from its use. The maturing of American police, and its elevation to a specialized, professional, coherent and reliable security body, was closely interconnected with the adoption of the telephone as its primary instrument of communication (Leonard 1938:introduction).

4.3 Early Uses of Wiretapping

There is an important sense in which the rate of adoption of telephony by British and American law enforcement and security agencies for internal communication and coordination purposes is not indicative of the extent to which they made full use of the system. What ought to be examined instead is the desire and ability of these agencies to expand their policing function into the realm of the telephone network. As telephony became widespread, reaching businesses, offices, government departments and schools, entering people's houses and appearing in public places, it gradually encompassed—and eventually channeled—the everyday functions of society. These functions were both mainstream and radical, legal and illegal. It was, therefore, not long before law enforcement and security agencies were drawn into assuming the responsibility of supervising and maintaining law and order in the virtual domain of the electric communications network. The vehicle of this newly assumed responsibility was the practice of wiretapping.

As a concept and a practice, the interception of voice communications dates back to the beginnings of telephony (Spindel 1968, p. 23). As is usually the case with new and untested communications technologies, the advanced interconnectivity facilitated by the telephone was seen as a threat, not only to the traditional social boundaries of class, but also the division between the public and private spheres of existence. The active concern of telephone customers about wiretapping can be observed in the numerous early attempts by entrepreneurs to increase the information security of telephone users through the invention of various gadgets that could be attached onto the telephone appliance. On both sides of the Atlantic, secrecy switches, jammers, voice scramblers, frequency changers and other such devices were produced and sold in large quantities. That was often done without the approval of telephone companies, which saw them as attempts to alter and distort the hardwired properties of the network by subverting the authority of main switching stations (Martin 1991, 20ff). In 1881, the opinion that customer control over information security would only be achieved when every user would "be his [or her] own operator" (Martin 1991, p. 145) was repeatedly expressed in the American press. Eventually, the telephone companies managed to retain control of the network, either through legal or administrative means. According to the Canadian sociologist Michéle Martin,

> had the telephone companies encouraged the development of secrecy devices attached to the telephone apparatus, privacy in the telephone system would have been entirely controlled by the users, and the structure of the system might have developed in a quite different manner (Martin 1991, pp. 21–22).

The importance of wiretapping, as well as of information security, was aptly demonstrated in warfare. During World War I, the existence of thousands of miles of telephone cables that crossed the battlefields of Europe, prompted the formation of numerous teams of undercover operatives on all sides of the conflict, who wandered around the continent attempting to penetrate enemy communication lines (Wingfield 1984, p. 18). Rival militaries competed in efforts to wiretap telephone cables belonging to the enemy. At the same time, they strove to protect their own communications through the use of various devices that enhanced information security. Undercover wiretapping posts set up by enemy forces were regularly uncovered in Britain, the United States and Germany throughout the duration of the conflict (Wingfield 1984, p. 19; Robertson 1947, p. 118). The paramount significance attributed by the British to telecommunications as a warfare-enabling system was illustrated by the *HMS Telconia* incident, which had an immense impact on the course and outcome of the War. One technical expert reports that

> [o]n the very first day of the war, 5 August 1914, the British sent the cable ship *Telconia* out to the North Sea with orders to grapple for, lift and cut Germany's transatlantic cables. With this operation successfully completed, Germany was forced to communicate with distant areas of the world either by radio or by using cables controlled by her enemies, [which were of course promptly and regularly intercepted] (Wingfield 1984, p. 19).

In addition to intercepting enemy communications, British and American intelligence agencies exercised extensive surveillance of domestic telegraph, telephone and mail exchanges throughout the War. This was done for purposes of monitoring, as well censoring, internal communications (Porter 1987, p. 179; Sampson 1973, p. 35).

It would be accurate to claim that the average telephone customer was familiar with the concept of information security long before World War I. It was known to early subscribers that conversations on the telephone network were closely supervised by the companies themselves, most of which adhered to strict policy rules regarding the use of "vulgar or obscene language" over the telephone (Marvin 1988, pp. 88, 89). In 1907, Canadian newspaper *The Ontario Globe*, asserted that the Bell Telephone Company had in its possession "the machinery for a system of espionage more than Russian in its perfection" (quoted in Martin 1991, p. 145).

In the United States, the first known incident of systematic wiretapping by a law enforcement agency was recorded in New York City in 1894. The incident was initially denied by officials in the police, the New York Telephone Company and Western Union Telephone Company. But the latter admitted in 1916 that the New York Police Department had been engaged in extensive wiretapping of telephones belonging to Catholic Church institutions and their members, in an attempt to collect evidence on suspected large-scale charity fraud. The project had been conducted with the full knowledge of senior Police officials, the telephone companies and the city's mayor, John P. Mitchell. What is more, the wiretappings appeared to have been conducted in breach of the relevant New York state laws, which explicitly outlawed the practice of telephone surveillance. During the 1916 Congregational investigation hearings, which followed the revelation of the scandal, a number of police officers testified that they had overheard countless personal conversations between lawyers and their clients, doctors and their patients, as well as between husbands and wives. The hearings were abandoned with the United States' entry into World War I and no disciplinary action was ever taken (Spindel 1968, p. 23; Gill 1994, pp., 4–5; Dash et al. 1959, 24ff; Diffie and Landau 1998, p. 155). Although few allegations of law enforcement wiretappings saw the light of publicity, it appears that the New York City Police Department resorted to illegal telephone surveillance on numerous instances in the years between 1892 and 1938 (Dash et al. 1959, pp. 26, 35–36). Police agencies in other American states, including Rhode Island, Connecticut and California, resorted to similar practices during those years (Dash et al. 1959, p. 162; Marvin 1988, p. 68).

Gradually, wiretapping began to be increasingly used for purposes of political policing. Since prior to the turn of the century, police officers in the city of New York had begun to wiretap the offices of a number of trade union organizations in the city (Dash et al. 1959, p. 27). A few years later, the then primitive technology of wiretapping was "promiscuously used" (Ungar 1975, p. 443) by the FBI. In Britain, the systematic wiretapping of the county's trade unions dates back at least to the 1926 General Strike, when the telephonic communications of the leader of the Transport Workers Union, Ernest Bevin, was targeted by the Special Branch (Fitzgerald and Leopold 1987, p. 27). The telephones of the Communist Party of Great Britain had begun to be systematically monitored shortly before the 1920s, in what some claim to have been the country's longest-running telephone surveillance operation (Andrew 1985, p. 368).

4.4 The Legal Pretexts of Wiretapping

Until the end of the twentieth century, the overwhelming majority of state-sponsored wiretapping operations in Britain were carried out by three government agencies, namely MI5, the London Metropolitan Police, and Customs and Excise (Dash et al. 1959, p. 290). From the end of the nineteenth century, and until the late 1930s, no specific permission or authorization was required for wiretapping. Instead, arrangements were made directly between law enforcement or security officers and General Post Office officials. More often than not, such arrangements included the interception of telephone calls by General Post Office employees on behalf of the police or MI5. The latter acted in the understanding that such practices were not in violation of British law (Lambert 1986, p. 208; Bunyan 1976, p. 197; Dash et al. 1959, pp. 288–289).

There was, however, an apparent inconsistency between the practice of mail and telephone interception: the former had required written authorization by the Home Secretary ever since 1663, while the latter was carried out without any ministerial oversight. It was in 1937 when this discrepancy was finally eradicated through a Home Office circular requiring the Home Secretary or his/her substitutes to authorize wiretapping operations by government agencies (Bunyan 1976, p. 39; Rolph 1973, p. 395; Dash et al. 1959, pp. 288–289; Lambert 1986, p. 208; Fitzgerald and Leopold 1987, pp. 56, 114). The procedure remained largely unchallenged throughout the Cold War, when telephone surveillance was routinely used for counterintelligence purposes, namely to ensure that no one who is known to be a member of the Communist Party or to be associated with it in such a way as to raise legitimate doubts about his or her reliability is employed in connection with work, the nature of which is vital to the security of the state (MI5 report from 1955, quoted in Dash et al. 1959, p. 291).

But in 1957, when the Metropolitan Police divulged outside of a Court of Justice the transcripts of an intercepted telephone conversation, British Prime Minister Harold Macmillan ordered the formation of an inquiring Committee of Privy Councilors, which later became known as the Birkett Committee (Command 283 1957). In its findings report, the Committee candidly declared that the origins of the government's authority to intercept telephonic communications were "obscure" (Command 283 1957, Sect. 9). It was so, said the findings report, because, although the practice of communications interception by state authorities had been systematically exercised and legally acknowledged since the seventeenth century, nowhere was there to be found an actual legal basis for it. In other words, "nowhere was statutory authority for the practice actually granted to the government" (Gill 1994, p. 164), which meant that the state appeared to possess no legal power to wiretap the telephone conversations of its citizens (Rolph 1973, p. 395; Fitzgerald and Leopold 1987, p. 112; Bunyan 1976, 197ff; Dash et al. 1959, pp. 288–289). At the same time, however, the Committee could find no legal statements that expressly outlawed the interception of communications by government agencies. The Committee was, therefore, compelled to conclude that, in the absence of a lucid authoritative statement outlawing communications interception by the state, the practice was to be considered lawful (Lambert 1986, p. 210; Bunyan 1976, p. 198). Furthermore, the lengthy historical practice of communications interception was considered to imply that "the power must have been vested in the Sovereign prior to the beginning of legal memory" (Fitzgerald and Leopold 1987, p. 113) and thus no explicit statutory authorization was required (Gill 1994, p. 164). That view echoed a 1952 statement by the then Home Secretary, Sir David Fyfe, who claimed that telephone interception was "a power which has been used by every government, of whatever political persuasion, since the telephone was invented, and is a Prerogative power" (quoted and Fitzgerald and Leopold 1987, p. 115).

In 1979, the legality of wiretapping was challenged again, this time by a James Malone, an antique dealer who was charged by the London Metropolitan Police with handling stolen goods. Malone was presented in court with transcripts of a number of his telephone conversations with burglars, which had been intercepted by Post Office Telecommunications employees. The objections of Malone's legal team were overruled by the Vice Chancellor of the Chancery Division, on the grounds that: (a) there was no basis for the right to privacy in English law; (b) the right of property could not be upheld in the case of telephone conversations; (c) the code of the European Convention for the Protection of Human Rights and Fundamental Freedoms was not enforceable in English courts; and (d) the absence of statutory dictums affirming the legality of wiretapping did not render the practice illegal (Gill 1994, p. 164).

In the United States, the comparatively speedy growth of the telephone generated public discussions about the phenomenon of wiretapping from quite an early stage.

For instance, having already outlawed telegraph wiretapping as early as 1861, the California legislature criminalized the practice of intercepting telephone communications in 1905 (Joy and Wright 1974, p. 253; Dash et al. 1959, p. 161; Westin 1962, p. 125). By 1909, a number of other states had followed suit, including that of Washington, whose statute explicitly declared that

every person [...] who shall intercept, read or in any manner interrupt or delay the sending of a message over any telegraph or telephone line [...] shall be guilty of a misdemeanour (quoted in Murphy 1965, p. 112).

The federal structure of the United States has given rise to a colorful mosaic of legislative interpretations of wiretapping, with almost as many versions of them as there are states: at various stages of the twentieth century, a number of states, including New York, Maryland, Massachusetts, Nevada and Oregon, introduced laws permitting local law enforcement agencies to wiretap with a court order. During the Cold War, approximately 38 states had statutes forbidding wiretapping by all persons, though most of them did not interpret their laws as applying specifically to the police, as was the case in Illinois and Pennsylvania. Several states, such as Louisiana, had, at various periods, laws permitting law enforcement officers to wiretap without a court order. Finally, many states allowed the use in court of telephone conversations obtained by police without court approval, while some, such as Pennsylvania, Illinois and Texas, explicitly forbade such evidence from being used in criminal trials (Westin 1962, p. 125, 153; Maguire 1959, p. 203; Murphy 1965, 149ff; Lapidus 1974, p. 45).

At the federal level, however, specific legislation regulating wiretapping was completely non-existent until 1968 (Carr 1998, Sect. 2.4; Lapidus 1974, p. 11). Prior to America's independence from Britain, the country's judicial system followed English common law. Following secession, the absence of organized law enforcement from non-urban areas meant that most instances of eavesdropping or mail-tampering were directly handled by victims (Friedman 1973, p. 254; Seipp 1977, p. 11; Flaherty 1989, p. 89). In 1878, the United States Supreme Court forbade the interception of first-class mail by law enforcement agencies without a court order (Diffie and Landau 1998, p. 129). As this ruling was not extended to cover telephone surveillance, federal law enforcement and security agencies engaged for many decades in extensive and unrestricted wiretapping that was practiced across the nation (Jeffreys 1995, pp. 199–200).

Paradoxically, the first significant challenge of that state of affairs was issued, not by a politician or a human-rights activist, but by a bootlegger. In 1928, crime baron Roy Olmstead was the first American citizen to be convicted primarily on evidence derived from communications interception. Olmstead took his case to the United States Court of Appeals and from there to the Supreme Court, claiming that the interception of his telephone conversations, as well as the divulgence of the transcripts of those conversations in court, infringed on his personal rights, and was thus antithetical to the fourth Amendment of the Constitution. By the narrowest possible majority of 5 to 4, the Supreme Court ultimately upheld Olmstead's conviction (Maguire 1959, p. 200; Overstreet and Overstreet 1969, p. 127; Lapidus 1974, p. 16;

Diffie and Landau 1998, 131ff; Murphy 1965, p. 112; Westin 1962, p. 124). The most controversial element of the decision was that the court had accepted the transcripts of Olmstead's telephone discussions as criminal evidence, even though the laws of the state of Washington—where the defendant lived and where he was arrested—explicitly outlawed the practice of wiretapping. Delivering the opinion of the Court, Supreme Court Chief Justice William H. Taft—who later became President of the United States—said that, although Washington state legislation forbade telephone interception,

> this statute does not declare that evidence obtained by such interception shall be inadmissible [. . .]. Whether the State of Washington may prosecute and punish several officers violating this law and those [individuals] whose messages were intercepted may sue them civilly is not before us (quoted in Murphy 1965, p. 112).

Harvard Law School professor John M. Maguire noted in 1959 that "never at any moment in its whole history has [a wiretapping case] commanded the approval of all members of the Supreme Court" (Maguire 1959, p. 200). That was true during the Olmstead case as it is true today, with only a handful of exceptions. Yet, the *Olmstead v. United States* precedent served to render federal-sponsored wiretapping constitutionally compatible.

Six years after *Olmstead v. United States* came the 1934 Federal Communications Act, which provided a basic regulatory framework for the interception of communications. Section 605 of the Act stated that "interception and divulgence" of telephone conversations was prohibited. However, those who were in favor of the imposition of tighter control on federal wiretapping practices were soon to be disappointed: it gradually emerged that courts throughout the nation were interpreting Section 605 to mean that a violation occurred, not upon the practice of interception, but upon the disclosure of the conversation to outsiders (Lapidus 1974, p. 11). It has been claimed, though not supported by evidence, that the United States Department of Justice—the government body behind the creation of Section 605— deliberately left a legal window open for federal law enforcement agencies to continue to wiretap without fearing prosecution (Lapidus 1974). Regardless of the motives behind Section 605, the fact remains that it had minimal—if any—impact on the wiretapping routines of the FBI and other federal agencies (Murphy 1965, p. 137; Gill 1994, p. 13; Diffie and Landau 1998, p. 157; Carr 1998, Sect. 1.10).

The perceived defects of Section 605 vexed one of the few influential adversaries of wiretapping, United States Attorney General Robert H. Jackson. In late February 1940, based on a recent Supreme Court ruling that declared wiretapping illegal, Jackson issued an authoritative directive instructing all United States attorneys to refrain from initiating any prosecution that included evidence arising from intercepted telephone conversations, regardless of whether that evidence was divulged in court. On 21 May 1940, acting under pressure by the FBI and other federal security and intelligence agencies, and with the very real possibility of American military involvement in Europe on the doorstep, President Franklin D. Roosevelt sent Jackson a confidential directive commanding the exception from his directive of cases relating to national security, namely cases involving "persons suspected of subversive activities against the government of the United States, including suspected spies', and the limitation of such cases "insofar as possible to aliens" (quoted in Wise 1976, p. 98; see also Murphy 1965, pp. 135–136).

Following the end of World War II, the controversial activities of the House Un-American Activities Committee sparked within government circles a relatively comprehensive discussion on wiretapping and the right of telephone users to information security. For different reasons, participants on all sides of the political spectrum consented that Section 605 was, in the words of the then-United States Attorney General Nicholas Katzenbach, "the worst of all possible solutions" (quoted in Wise 1976, p. 22). From this consent, substantiated by a number of successful legal challenges—including *Katz v. United States* (1967) and *Berger v. United States* (1967)—emerged Title III of the 1968 Omnibus Crime Control and Safe Streets Act, which has defined the regulatory framework of communications interception in the United States ever since. Title III of the Act was overwhelmingly approved in June of that year by Senate and Representative majorities. It constituted the inaugural official sanctioning of federal wiretapping in the nation's legal history. While outlawing eavesdropping of all kinds by private individuals, Title III limited the use of authorized wiretaps to the investigation of specific types of crime—such as kidnapping, gambling, counterfeiting, murder, etc.—and barred the use of unlawfully intercepted telephonic communications as evidence in court. Furthermore, it specified the procedure of warrant authorization required for wiretapping operations and limited their maximum duration to 30 days. Wiretapping without court authorization was also facilitated by Title III in two cases, namely during a declared 48-h emergency, and for national security purposes (Carr 1998, Sects. 3.143–144; Lapidus 1974, 1ff, 42; Gill 1994, p. 15; Diffie and Landau 1998, p. 171).

4.5 The Regulatory Dysfunction of Historical Wiretapping

It is difficult to estimate the precise number of government agencies that possessed and used wiretapping equipment in the twentieth century. In the United Kingdom, local police departments, the Scotland Yard, MI5, as well as Revenue and Customs, had the power to conduct their own domestic wiretapping investigations. There is no reason, however, why other agencies, such as those concerned with quality control, safety regulations, the nuclear industry, social care or pensions, would not have subscribed to the practice—though records are imprecise. In the United States it is known that, apart from local and federal law enforcement agencies, the Secret Service, the Internal Revenue Service and various branches of the Department of Justice all possess domestic communications-interception capabilities. With reference to the United States, it is believed that, by 1972, over 60 state agencies had access to telephone surveillance hardware (Lapidus 1974, p. 12).

In both the United Kingdom and the United States, the sheer number of state authorities, boards and agencies that possessed wiretapping capabilities hampered the strict executive control of the practice throughout the twentieth century. In the

United States, for instance, local FBI field offices usually incorporated telephone monitoring studios for use by local agents (Ungar 1975, p. 188). These so-called *radio rooms* were regulated and administered by the local field FBI offices, and thus information as to their use tended to remain in local files, unless there were specific reasons for them to be dispatched to the FBI headquarters in Virginia (Elliff 1972, p. 26). Consequently, it was almost impossible for the Director of the FBI, let alone the Attorney General, to become aware at any particular point in time of the exact extent of wiretapping exercised by the FBI throughout the country.

There were also further bureaucratic complexities, of which the United States is a good case in point: a single warrant permitting telephone surveillance in aid of a specific investigation of a suspect could potentially be used to cover, not only the main suspect, but all telephone users, organizations and institutions which may be connected to the suspect. Finally, British observers have claimed that, in areas such as Northern Ireland, law enforcement and intelligence bodies have, for decades, intercepted telephone communications in the absence of adequate ministerial authorization (Wingfield 1984, p. 28; Lapidus 1974, p. 6; Lambert 1986, pp. 212, 218). If all these claims are taken into account, then a picture begins to emerge of a state apparatus that often failed to keep track of its wiretapping operations as the use of the telephone spread in the twentieth century. Critics have claimed that, in the past, the relaxed internal monitoring system over communications interception gave government agencies "free rein in collecting wide-ranging domestic intelligence" (Diffie and Landau 1998, p. 162).

In Britain, the very concept of communications interception by the state has been based on the markedly secretive character of the state's executive branch (Robertson 1982, 22ff; Harden and Lewis 1988, Chap. 4). Consequently, although state agencies in possession of wiretapping hardware have been required to operate on telephone surveillance quotas—established and supervised solely by the Home Office Secretary—the extent to which ministerial approval has been consistently sought by British law enforcement agencies for communications-interception operations has been strongly disputed in the available literature (Gill 1994, p. 220). Patrick Fitzgerald and Mark Leopold, for instance, report that, during the Cold War, MI5 officers would usually instigate wiretapping operations upon suspicion and would proceed to apply for a warrant only if their suspicions were confirmed by the intercepted conversations (Fitzgerald and Leopold 1987, pp. 57–58).

Historically, the relative—complete, prior to 1986—absence of a clear legislative basis for state wiretapping in Britain has practically excluded Parliamentary control over the practice. State officials have often been notably forthright about these arrangements. Speaking in the House of Commons in April 1980, the British Home Secretary candidly confessed that

[t]he interception of communications is, by definition, a practice that depends for its effectiveness and value upon being carried out in secret and cannot, therefore, be subject to the normal processes of Parliamentary control (quoted in Lambert 1986, p. 210).

In the United States, commentators have recognized the "relative inability of Congress" to regulate state-sponsored communications interception as one of the most "interesting aspects" surrounding the issue (Murphy 1965, p. 157). The periodically resurfacing National Wiretapping Commission has suggested that

> despite the [Justice] Department's assertions to the contrary, there is an absence of well-defined procedures which would promote compliance with the statutory standards and permit meaningful congregational scrutiny of this extraordinary executive activity (quoted in Navasky and Lewin 1973, pp. 298–289).

Throughout the last century, the United States Department of Justice was repeatedly accused of being unable to establish an adequate system of control over wiretapping by law enforcement agencies. In 1972, under the pending shadow of the Watergate scandal, the United States Supreme Court decided by an 8 to 0 vote that a court order was required to wiretap the telephone apparatuses of American citizens who did not appear to have "significant connection with a foreign power, its agents, or agencies" (quoted in Lapidus 1974, p. 97). Shortly afterwards, the United States Attorney General Richard Kleindienst personally pledged to eliminate all illegal electronic surveillance operations. Yet, less than a fortnight later, Department of Justice officials reported to Congress that wiretapping of domestic radicals without apparent connections to foreign agencies was indeed continuing (Gill 1994, p. 16). A few months later, the exposure of the Watergate scandal caused Kleindienst to resign. During the Watergate investigations, former US Attorney General R. Clark stated:

> [r]eports by FBI agents on electronic surveillance had caused the Department [of Justice] deep embarrassment many times. Often we would go to court and say that there had been no electronic surveillance and then we would find out we had been wrong. Often you could not find out what was going on [...] frequently agents lost the facts (USSSCSGORIA 1976b, pp. 337–348).

This absence of executive controls over the wiretapping practices of American state agencies is often said to have its roots in the absence of a culture of accountability on the part of those agencies during the Cold War. An illustration of this can be seen in the insistence of the Director of the National Security Agency (NSA) in 1981 that the United States Congressional Committee on Governmental Operations forward him a copy of its investigative report entitled *The Government's Classification of Private Ideas* prior to its official issuance. In disbelief, the Committee pointed out to the NSA that "Congress does not submit its reports to Executive Branch agencies for preview" (United States House of Representatives 1987, pp. 21–22).

4.6 Wiretapping in Practice

The absence of substantial oversight of domestic communications interception effectively means that attempting to analyze the qualitative and quantitative aspects of government wiretapping by reference to its official, legal framework is exceedingly difficult and—ultimately—unproductive. The sensitivity of the practice has kept it largely concealed from popular scrutiny. Moreover, the foremost executive principle guiding government-sponsored wiretapping ever since the inception of the telephone has been that of maximum operational secrecy and, in some cases, political camouflage.

The actual extent of the use of wiretapping by state agencies in the twentieth century may eventually come to light. Throughout that period, there was little hope for an effective and open dialogue, since telephone surveillance was "barely acknowledged by the authorities" (Lambert 1986, p. 208). Even investigative reports, such as the 43-page account of the aforementioned Birkett Committee (Command 283 1957), arguably the most extensive official source of information on wiretapping in Britain, have been described by observers as "seriously incomplete" (Fitzgerald and Leopold 1987, p. 120). As is the case with many official figures, the validity of governmental estimations of wiretapping differ, often drastically, from estimates provided by independent observers. The former are thus heavily disputed. In 1954, for instance, Supreme Court Justice William Douglas asserted that "during 1952 there were in New York City alone at least 58,000 orders issued which allowed wiretapping—over 150 a day every day in the year" (Douglas 1954, p. 355). Yet, official figures of authorized wiretappings carried out during that same year in the entire state of New York claimed no more than 419 authorizations (Dash et al. 1959, p. 42). In Britain, claims made by the national press that there had been almost 2000 warrants issued to the London Metropolitan Police in 1971 and 1972 were rejected by the Minister for State, Mark Carlisle, as "ludicrously high'. Carlisle refused to disclose official figures in the name of the national interest (Bunyan 1976, p. 201; Fitzgerald and Leopold 1987, p. 130).

4.7   Considerations on the Actual Extent of Wiretapping

For decades, critics of government-sponsored communications interception have claimed that, during the Cold War, the absence of parliamentary control in domestic wiretapping encouraged, if not practically directed, law enforcement and security agencies to use their wiretapping capabilities far in excess of legal limitations (Navasky and Lewin 1973, pp. 298–289). More often than not, the misuse of wiretapping laws is said to have been achieved through institutionalized lying and deception, "denial and related arts" (Donner 1980, p. 24). In the United States, this was first practiced, and indeed perfected, under the supervision of J. Edgar Hoover, Director of the FBI for nearly half a century. For 34 consecutive years, from 1924 to 1957, Hoover's public attitude toward wiretapping Americans was highly negative. He often resorted to vocal remarks against the practice. In a 1931 Congressional hearing, for instance, he said:

[w]e have a very definite rule in the Bureau that any employee engaging in wiretapping will be dismissed from the service of the Bureau [...]. While it may not be illegal, I think it is unethical, and it is not permitted under the regulations by the Attorney General (quoted in Murphy 1965, p. 129).

And in 1939 he remarked:

> [w]hile I concede that the telephone tap is from time to time of limited value in the criminal investigation, I frankly believe that if a statute of this kind were enacted, the abuses therefrom would far outweigh the value which might accrue to law enforcement as a whole (quoted in Lapidus 1974, p. 67).

Behind the scenes, however, with Hoover's full knowledge and support, the FBI was already engaging in extensive wiretapping operations as early as 1931 (USSSCSGORIA 1976a, p. 12; Gill 1994, p. 8). As is evident in the—now declassified—*FBI Manual*, Hoover actively encouraged Bureau agents to use domestic communications-interception techniques to uncover perceived communist influence in areas such as "political activities, Negro question, youth matters, women's matters, farmers' matters, veterans' matters" (*FBI Manual* Sect 87, pp. 5–11, quoted in USSSCSGORIA 1976a, p. 363). Hoover's abuses have been described by critics as "monumental" (Kessler 1993, p. 375). Until his death, which found him in office, he wiretapped numerous members of Congress, political personalities of all convictions, Supreme Court Judges, campaigners, activists and entertainers (USSSCSGORIA 1976b, pp. 313–314, 345; Murphy 1965, p. 158). Many insist that, to cover his abuses, Hoover deceived Congress and the American public alike. Allegedly his most often-used technique was to have the majority of operating FBI wiretaps shut down for the day he presented his annual testimony to Congress, so that he would not have to lie about the number of active wiretaps. He would then have them resumed on the following day (Donner 1980, p. 24; Jeffreys 1995, p. 200; Diffie and Landau 1998, 271n; USSSCSGORIA 1976b, p. 302).

The outbreak of World War II gave Hoover the pretext to recognize publicly for the first time that changing world developments pointed to the utilization of wiretapping for purposes of preventing espionage and sabotage by aliens (Murphy 1965, pp. 137–138). Yet, along with suspected aliens, primary targets of wiretapping operations during World War II by American civilian and military intelligence agencies were—apart from the Roosevelts (Wise 1976, 98n)—nuclear scientists and scores of legal labor organizations, including

> the Congress of Industrial Organizations' Council and Marine Committee; the Tobacco, Agricultural, and Allied Workers of America; the International Longshoremen's and Warehousemen's Union; the National Maritime Union; the National Union of Marine Cooks and Stewards; and the United Public Workers of America (Diffie and Landau 1998, p. 138; see also Theoharis and Cox 1988, pp. 10, 438).

World War II was followed by the onset of the Cold War and by an almost complete lifting of all regulatory frameworks of wiretapping. Researchers assert that, in 1948, the New York County Criminal Courts Bar reported that the New York police had reached the stage of employing wiretaps for the most minor of offences (Dash et al. 1959, p. 33). On 4 March 1949, Department of Justice employee Judith Coplon was arrested in New York City by FBI agents as she was about to hand classified FBI documents to Valentin Gubitchev, a Soviet diplomat stationed at the United Nations mission in New York. Following a request by Coplon's defense lawyers, the content of the classified documents was publicly disclosed during her trial, to the detriment of the FBI. It revealed details of an illegal FBI wiretap operation targeting pro-New Deal academics, as well as Hollywood entertainers who had publicly rebuked the activities of the House Un-American Activities Committee and had expressed support for Henry Wallace's 1948 Progressive Party campaign (Theoharis 1978, p. 100).

It was indeed during the tense period of McCarthyism that the practice—as well as the fear of—systematic telephone surveillance spread, reaching virtually every sector of state and civil society in America, down to the local level. Numerous investigations conducted by newspapers, state committees and Congress raised strong suspicions of illegal wiretapping by local law enforcement officials throughout the country (Brenton 1964, p. 163). Wiretap orders for purposes of political surveillance against individuals and groups were "obtained in quantity and in blank" (Spindel 1968, pp. 43, 202; see also Donner 1980, p. 244, and Westin 1962, p. 152), in direct violation of federal and state laws. In the state of New York, for instance, official figures showed approximately 3000 wiretap orders for the period between 1950 and 1955. However, in what is the most detailed study of wiretapping ever undertaken in the United States, a team of researchers working under the direction of Philadelphia attorney Samuel Dash—who later served as chief counsel for the Senate Watergate Committee during the Watergate scandal—concluded that, with the addition of unauthorized wiretaps operated by New York state police, the figure for that period rose to somewhere between 96,000 and 174,000 (Dash et al. 1959, p. 68). It also seems probable that extensive communications-interception operations were conducted against state officials, especially those working in defense posts. For instance,

> [n]ewsman Ben H. Bagdikian, on the Washington scene for a number of years, stated in a *Saturday Evening Post* article on news managing that a considerable number of Pentagon officials take it for granted that their offices are bugged. And 'almost every defence correspondent I talked to', he added, 'assumed his telephone, office and home are tapped by some government agency' (Brenton 1964, p. 164).

The 1950s also witnessed the most extensive illegal communications interception operation by a law enforcement agency ever exposed in the United States. In February 1955, what was in effect an illegal wiretapping station maintained and operated by New York Police Department officers and a number of private detectives, as well as two New York Telephone Company personnel, was uncovered in downtown New York City. Through it, police officers were able to monitor more than 125,000 telephone lines on the City's East Side, including the Manhattan business sector, numerous diplomatic missions and consulates, and even the telephone network of the United Nations headquarters. Four people were arrested, all of whom denied operating on behalf of the FBI or other US government agencies (Spindel 1968, p. 26, 105ff; Westin 1962, p. 152).

Researchers claim that the divisive political landscape of America during the 1960s was met by a further wave of illegal wiretapping operations. These wiretaps were used by law enforcement and security agencies either as means of neutralizing targeted political groups, or simply gathering information (Donner 1980, p. 245;

Ungar 1975, p. 449). Often, the *possibility* of communist infiltration of dissident political groups was seen as sufficient to justify warranted or unwarranted telephone surveillance, which often lasted for many months, and in some cases even years (Donner 1980, p. 244). This justification was used to target the civil rights movement and in particular its leader, the social reformer Dr. Martin Luther King, Jr. The latter was particularly targeted, even though he was fiercely critical of communism and any form of political violence against the state. Initially, wiretaps were installed on King's home telephone line and two of his office telephones. Later, and with the full knowledge of the director of the FBI, the president of the United States, and the Attorney General, who authorized the wiretaps, telephone surveillance was extended to cover "King's advisers, hosts, visitors, and associates, fifteen hotel rooms where he stayed, and [...] even his [church] pulpit" (Donner 1980, p. 244; see also Diffie and Landau 1998, p. 142). Other civil rights campaigners, including Elijah Muhammad and Malcolm X, were also targeted. In one particular incident, an official of the Nation of Islam with a blank criminal record was constantly wiretapped for no fewer than 8 years, but never prosecuted (USSSCSGORIA 1976a, p. 63). In June 1971, Supreme Court Justice William Douglas stated that, on average, unwarranted "national security" wiretaps lasted for 78 to 209 days, as opposed to an average of 13 days for other authorized wiretaps (USSSCSGORIA 1976a, p. 101).

All wiretapping scandals that were exposed during the tumultuous decade of the 1970s were inevitably diminished by the shadow of Watergate. President Richard Nixon often liked to stress his supposed curbing on unauthorized wiretapping by state agencies. In March 1971, for instance, he reported to the Senate:

> in the two years that we have been in office, now get this number, the total number of taps for national security purposes by the FBI, and I know because I look not at the information but at the decisions that are made—the total number of taps is less, has been less, than fifty a year (quoted in Gill 1994, p. 22).

In 1973, on the very first full day of his second administration as president, Nixon assured reporters that his entrusted Attorney General, John Mitchell—later convicted and imprisoned over the Watergate scandal—would govern telephone surveillance "with an iron hand" (White 1975, p. 125). And yet, the first illegal wiretappings by Nixon and his collaborators were to begin less than 4 months after that comment. Along with Watergate, the Nixon administration was responsible for another sinister scheme involving wiretapping, often called *the Huston Plan*, after its designer, Nixon's aide Thomas Huston. In a detailed report to the President, Huston warned of a

> connection between domestic unrest and foreign movements and proposed a plan in which the resources of the CIA, the [Defense Intelligence Agency], the FBI and the NSA were to be pooled to fight domestic unrest. The NSA—contrary to law—would intercept communications of American citizens using international facilities. Rules regarding mail interception, electronic surveillance, and surreptitious entry would be relaxed [...]. That the Huston Plan came within a hairsbreadth of being national policy shocked the nation when it was revealed several years later (Diffie and Landau 1998, pp. 145–146)

After Nixon's resignation, the Church Committee, a body of United States Senators who conducted a detailed, year-long investigation of past activities of American intelligence and law enforcement agencies, characteristically reported that the nature, extent and pernicious nature of domestic intelligence operations "threaten to undermine our democratic society and fundamentally alter its nature" (USSSCSGORIA 1976a, p. 1).

In Britain there has been comparatively little evidence of actual wiretapping misuse by the authorities during the twentieth century. Yet scholars claim that the reasons have less to do with higher legal standards upheld by security and law enforcement agencies and more with the traditional secrecy of the British state, as well as with the equally traditional institutional intimacy between law enforcement and security agencies and Post Office Telecommunications, Britain's foremost telephone service provider—which is discussed later in this book. In one instance in 1985, Member of Parliament John McWilliam, a former telecommunications engineer, stated in Parliament that "[i]t is wrong to say that interception has not happened without warrants. In addition to officially authorized taps certain official tappings do not require warrants. That derives from an institutional relationship between the police, the CID or the Special Branch, and the Post Office—subsequently British Telecom" (quoted in Fitzgerald and Leopold 1987, p. 15). Statements such as this directly contradicted official documents by successive British governments in the twentieth century, which described state-sponsored wiretapping as a marginal activity that law enforcement agencies strove to avoid as much as possible, but ultimately resorted to it "from time to time" (Gill 1994, p. 168; Command 283 1957, various pages throughout).

In theory, British state agencies with access to wiretapping hardware have traditionally operated according to quotas, which have allowed them to conduct a certain amount of wiretapping operations every year (Command 108 1987, Sects. 53–54). Official tables were published by British administrations throughout the twentieth century, up until 1989. These tables contain the interception of communications of all kinds, authorized by the Home secretary and—to a lesser extent—the Scottish as well as the Foreign Secretaries for England, Scotland and Wales, from 1937 to 1989 (Command 7873 1980; Command 9843 1985; Gill 1994, p. 168). Scholars have indicated that the statistics contained in these tables remained "surprisingly stable" (Lambert 1986, p. 212), at under 600 a year, while from 1969 to 1989 wiretapping authorizations never exceeded a total of 470 (Gill 1994, p. 168). What prompts the skepticism of observers is the fact that during those two decades the number of telephone lines in Britain increased by 250 percent, from 8 to 20 million (Gill 1994). According to Fitzgerald and Leopold,

> the number of warrants quoted in the White Paper bears little relation to the number of phone calls tapped by the authorities. Far more tapping goes on than has ever been admitted. [Besides the fact that] the numbers of warrants signed by the Northern Ireland Secretary, which may include taps in mainland Britain, have never been released [...], the [official] figures relate to those warrants *signed* in a particular year, not to those *in force* during that year. Warrants only stay in force for a matter of months, but they may be renewed

indefinitely [—the so-called *blanket tappings*]. Many of those signed in previous years may, therefore, still be operative (Fitzgerald and Leopold 1987, p. 14).

In addition, scholars argue, it is possible that, similarly to the United States, a single warrant permitting telephone surveillance in aid of a specific investigation of a suspect can potentially be used to cover not only the target suspect himself or herself, but all telephone users, organizations and institutions that may be connected with him or her (Fitzgerald and Leopold 1987, p. 15; Gill 1994, pp. 168–169). Finally, it has repeatedly been alleged in Parliamentary debates—though never substantiated with evidence—that "[t]he problem is not phone taps authorized by the Secretary of State, it is the very much larger number of phone taps that are made without any application for an authorization', for instance by municipal police forces (Member of Parliament Ian Mikardo in 1985, quoted in Fitzgerald and Leopold 1987, p. 15).

In the alleged absence of concrete and trustworthy data, researchers have regularly resorted to employing logical inference as a methodological tool. Technical expert John Wingfield, for instance, claimed in 1984 that the less than 500 wiretapping orders—both domestic and international—issued by the British government every year did not justify the sheer size of units that were solely concerned with communications interception. In 1982, he continued, it was reported that one state-sponsored telephone surveillance site in London employed over 100 technicians and operated on an annual budget of £1.25 million (Wingfield 1984, p. 28). Others furthered this position by stating that, by the late 1980s, the Government Communications Headquarters (GCHQ) employed 11,500 full time staff, including 7000 civilians and 4500 service personnel (Fitzgerald and Leopold 1987, p. 48; Norton-Taylor 1990, p. 53). To that, Professor Peter Gill insisted in 1994, one should add the wiretapping facilities maintained and operated by British Telecom—the privatized reincarnation of Post Office Telecommunications—itself. By the mid-1990s, the company had

> 60 full-time tapping operatives [as well as] 80 higher grade engineers maintaining and servicing tapped lines. At the BT center in Gresham Street there [were] more than 1000 lines plus switching mechanisms [...] deal[ing] with many thousands of incoming intercepts. [The center was staffed by] 100 transcribers while [a]dditional transcribers work[ed] at MI5 on the material directed there (Gill 1994, p. 167).

Clearly, critics conclude, the *potential* indicated by the technical and administrative capacity of British governmental agencies has been significantly greater than the official disclosures of the extent of wiretapping operations in the country in the 1900s.

4.8   The Industry-State Telecommunications Interface

The practice of communications interception is politically sensitive, not only because it evades privacy, but also because it represents one of the few functions of security and law enforcement agencies that involve the co-operation and approval of non-governmental agencies—namely the privately owned telephone service providers. Historically, it has not been strictly necessary for security investigators to rely on the assistance of the telephone companies for the execution of wiretapping operations. In the days when most telephone cables could still be visibly traced in public areas, basic equipment such as a pair of pliers, a wire and a pair of headphones could be used to wiretap a target line from the local telephone junction box—the latter being the most accessible interception point in the local network. From the standpoint of security, however, such operations have usually been problematic. In the words of a Post Office Telecommunications engineering report, wiretaps installed at local telephone junctions "present[ed] formidable technical and practical problems [and were] too open to public view and too liable to discovery by Post Office engineers on normal duties" (Post Office Engineering Union 1980, p. 20).

Additionally, the acceleration in the development of telephone technology, especially after World War II, made the task of wiretapping immensely more complex over the years, even prior to digitization. Identifying the telephone number of a target suspect is based on information that is available even to the average subscriber through the *Telephone Directory*. But as telephone lines were usually strung together in bunches of up to 100, identifying the particular telephone cable and pair at the junction box was practically impossible without access to expert knowledge and specialized technical data that were available only to the telephone company, and used by linemen on the job (Brenton 1964, p. 160; Dash et al. 1959, p. 66, 293; Lapidus 1974, p. 124). These complex technical problems could only be eliminated with the direct and substantial co-operation of the telephone carrier. Ideally, therefore, wiretaps were installed and maintained on the main distribution frame at the local telephone exchange or at the central telephone offices.

The necessary link ensuring the smooth and efficient co-operation between state security or law enforcement agencies and TSPs rested on intimate institutional relations—that is, reliable "institutional arrangements" (Bunyan 1976, p. 202) between those two actors. In countries such as Britain, France or Italy, where telephone carriers were owned and directly controlled by the state for many decades, those institutions formed close functional relationships, which played a major role in the systematic exercise of communications interception throughout the twentieth century (Plate and Darvi 1981; Fitzgerald and Leopold 1987, p. 9).

Scholars note that, in both Britain and the United States, getting telephone company assistance for authorized wiretapping operations has historically been "easier than getting a dial tone" (Gill 1994, p. 23; see also Wise 1976, 37n). Upon display of a valid communications interception warrant, telephone companies would provide agency representatives with either a leased line transferring targeted telephone conversations to a monitoring location, or with actual wiretapping instruments, which could be used on telephone company premises. The company was eventually compensated for its expenses by the agencies concerned (Carr 1998, Sects. 127–128; Diffie and Landau 1998, p. 172; Lapidus 1974, p. 124). But unauthorized wiretapping, which, critics claim, has historically been considerably more extensive than authorized wiretapping by state agencies, depends almost completely on telephone company assistance (Dash et al. 1959, pp. 72, 293).

In Britain, the history of the institutional intimacy between organized law enforcement agencies and TSPs was initiated on the very year of the establishment of the London Metropolitan Police, in 1829. At that time, the Confidential Enquiry Branch was formed and assigned the task of supervising the content of suspicious correspondence distributed by the General Post Office (Pertrow 1994, p. 264). During the late nineteenth century, under the directions of the Metropolitan Police Assistant Commissioner, the Branch utilized its position inside the General Post Office to initiate an extensive and lengthy operation against the increasing use of the mail and telephone for gambling and prostitution. In an instance that is illustrative of the complex nature of cross-departmental negotiation, the General Post Office refused to co-operate with the London Metropolitan Police by forbidding its officers to intercept the communications of British citizens. The General Post Office Solicitor, Sir Richard Hunter, protested that he was a carrier of messages, not "a censor of public morals" (quoted in Pertrow 1994, p. 266). In a fierce response, Harry B. Simpson, head of the Criminal Department of the Home Office, accused Sir Hunter of being an employee of

> a carrier who actually ascertains, or might with ordinary intelligence easily ascertain, that his services are being used for an illegal purpose [and he, therefore, could not avoid] criminal responsibility (Pertrow 1994).

But the uncooperative attitude of the General Post Office continued. In 1897, the Postmaster General, Henry Fitzalan-Howard, insisted that he could not assist the police unless he was empowered by Parliament (Pertrow 1994). In 1906 his successor, Postmaster General Sydney Buxton, insisted that sufficient policing of the telephone network implied nothing less of "a general censorship" of all conversations, something which he found unacceptable (Pertrow 1994, p. 270). In 1910, the Home Office was able to completely succumb the resistance of the General Post Office, by resorting to the appointment of Herbert Samuel, a public administrator who was favorable to the communications interception mandates of the Metropolitan Police, to the post of Postmaster General (Pertrow 1994, p. 268).

During the same year, the newly-founded MI5 was busy establishing "valuable contacts" (Fitzgerald and Leopold 1987, p. 26) with the General Post Office, and arranging the facilitation of postal and telephonic surveillance. Walter Thompson, a former inspector who was involved in the formation of the organization, wrote during that time that "whereas the ordinary detective can always feel that the Law stands firmly behind him, the Special more often than not lacks this advantage" (quoted in Porter 1987, p. 176). Until 1937, this peculiar institutional sanctioning of illegality governed the relationship between British security and law enforcement agencies and the General Post Office with regard to telephone surveillance. Up until that year, not only did the General Post Office directly negotiate with MI5, the Special Branch and the police, permitting the interception of telephone communications, but also performed such tasks on behalf of those agencies (Rolph 1973, p. 395; Bunyan 1976, p. 197; Fitzgerald and Leopold 1987, pp. 113–114):

[it] maintained a separate Telephone Interception Unit, which liaised directly with MI5's counter-espionage division and transcribed conversations on its behalf from recordings made on wax-coated cylinders (Fitzgerald and Leopold 1987, pp. 64–65).

Following the end of World War II, the General Post Office transcribers were permanently absorbed by MI5 and placed in the agency's Section A3—MI5's administrative and technical division, later renamed A2A (West 1983, pp. 18–19).

After 1937, when the requirement of a warrant for intercepting telephone communications became a standardized process, every General Post Office official had the right to refuse co-operation with the authorities unless—and sometimes even if— a valid warrant was displayed. The extent to which that right was put to use is, however, questionable, considering that the organic institutional intimacy between security agencies and the General Post Office had already been functional for almost three decades at that point. In other words, it is very likely that the majority of law enforcement or security officers "inclined simply to ignore the warrant system and make a direct approach to the [General Post Office], via the area head of the General Post Office Investigation Division" (Fitzgerald and Leopold 1987, p. 58). For most of the twentieth century, the General Post Office Investigation Division was the principal point of interface between the General Post Office and law enforcement and security agencies in the United Kingdom. Although it operated under the control of the General Post Office, its members were usually recruited from the ranks of former police or security officers, while a minority consisted of telephone technicians and engineers. After the privatization of British Telecom in 1981, the group was renamed to British Telecom Investigation Department and staffed by approximately 300 employees.

One area of co-operation that was totally unaffected by the 1937 change in legislation, was the monitoring and recording of telephone conversations conducted by General Post Office employees on behalf of security or law enforcement agencies. There were numerous reports of instances of Post Office Telecommunications local exchange buildings that incorporated monitoring stations staffed by company employees who routinely intercepted, recorded and even transcribed telephone conversations on behalf of security or law enforcement agencies (Bunyan 1976, 202; Fitzgerald and Leopold 1987, pp. 63–64). That was the case in 1952, when local councilors visiting the local British Telecom exchange in Slough, Berkshire, were shown a fully equipped and operational listening post by a fellow councilor, who was also an employee at the site (Bunyan 1976, p. 203). In 1979, during the case of *Malone v. Commissioner of Police for the Metropolis No2* (1979, A11 ER620) it was revealed that Post Office Telecommunications employees had intercepted and recorded telephone conversations upon the display of a warrant by the London Metropolitan Police (Lambert 1986, p. 209). These arrangements were persisting by the time of the appearance of a 1969 Home Office circular, which stated that

if the police are in urgent need of information which the Post Office may be able to furnish in connection with a serious criminal offence, the police officer in charge of the investigation should communicate with the Duty Officer, Post Office Investigation Division who will be ready to make any necessary inquiries (Home Office 1977, Sect. 1.70; see also Leigh 1975, 215ff).

This statement, which was seen at the time as diminishing the right of refusal by Post Office Telecommunications to collaborate with law enforcement agencies, was accompanied a few months later by a series of ministerial decisions regarding the relationship between Post Office Telecommunications and the Government Communications Headquarters (GCHQ). Namely, the government proceeded to announce the establishment of a newly integrated communications interception complex, created under the technical co-operation of both organizations. Technical improvements included the facilitation of immediate redirection by Post Office Telecommunications to GCHQ of targeted telephone conversations from anywhere in Britain (Fitzgerald and Leopold 1987, p. 68).

In 1981, along with the privatization of British Telecom, a new requirement for the company's employees was introduced: they all had to sign the Official Secrets Act, which, among other things, penalized employees for publicly disclosing information relating to wiretapping practices (Fitzgerald and Leopold 1987, p. 30). Here is how Fitzgerald and Leopold viewed at the time the way in which British Telecom's privatization impacted on its co-operation with law enforcement:

> [u]niquely for a private company, [British Telecom] now has a vital, integral role within the British intelligence complex. Should [British Telecom] ever do the unthinkable, and suddenly refuse to co-operate with official telephone tapping, the government has the safety net of a 'national security' clause, written into the privatization law [...]. Section 94 of the 1984 Telecommunications Act allows for a Secretary of State to issue [British Telecom] and any other telecom companies with such general or specific directives 'as appear to the Secretary of State to be requisite or expedient in the interests of national security'. The directive must be implemented, continues the section, irrespective of other obligations (to shareholders, for instance). Other clauses permit the Secretary to withhold the directive from Parliament 'in the interests of national security' and to reimburse the carriers for the cost of complying with the directive. The carriers are prevented from revealing the fact or substance of any such directive (Fitzgerald and Leopold 1987, p. 82).

In the United States there is little evidence of early involvement of security or law enforcement agencies in the everyday running of the Bell Telephone company. The closest example is the coercion imposed on the company in 1902 by the Canadian government to allow officers to search for irregularities in the records of telephone subscribers. It remains uncertain, however, whether records of customers in the United States were searched during that operation (Martin 1991, p. 44). The first recorded instance of a "loose arrangement" (Dash et al. 1959, p. 25) between a law enforcement agency and a telephone company occurred in 1895 in New York City, where the local carrier systematically assisted the police department in its wiretapping operations during a fraud investigation (Dash et al. 1959). But, on balance, there is an absence of significant evidence of intimate collaboration between law enforcement agencies and telephone companies during the early decades of the twentieth century. The subsequent evolution of that relationship may lead to the conclusion that it was primarily the personal connections between individual law enforcement and security officers and telephone company employees that provided isolated areas of functional overlapping.

It was during World War II that this overlay was refined and gradually institutionalized. In 1942, the Office of Naval Intelligence requested that the Record

Company of America (RCA) Global provide copies of all Japanese cable traffic to and from the island of Hawaii. Additionally, virtually all international telegraph messages carried by RCA Global, IT&T World Communications and Western Union International were intercepted by military intelligence officers throughout the duration of the War (Theoharis 1978, p. 120). The intimate institutional settings facilitated during those exercises permitted their continuation after the formal end of World War II. In August 1945, military intelligence expressed interest in retaining postwar access to messages of foreign governments carried by the aforementioned service providers. The companies accepted after having secured formal assurances by no other than the United States Attorney General Tom C. Clark and Defense Secretary James Forrestal that they would be legally protected. They were indeed protected, even though their consent to allow the interception of cable traffic was in direct violation of Section 605 of the 1934 Federal Communications Act (Theoharis 1978). The scheme, officially known as Operation SHAMROCK, was maintained until 15 March 1975, when it was revealed in the press in the context of the post-Watergate Congressional investigations. During the Cold War, this institutional intimacy proved invaluable for intelligence authorities. Through it, the NSA managed to obtain access to cable traffic carried by Western Union, RCA Global, and IT&T to and from selected domestic and international intelligence targets (USSSCSGORIA 1976a, p. 108).

During that time, the telephone companies' disclosure of telephone information to law enforcement agencies was rarely challenged under Section 605 (for instance, in *Brandon v. United States* in 1941 and *Bubis v. United States* in 1967), and rarely effectively (Carr 1998, Sects. 1.13–14). All possibilities for successfully challenging that activity were eventually diminished in 1970, with the amendment of Title III of the Omnibus Crime Control and Safe Streets Act. Under Section 2511(2)(a)(ii) of the amended Title, telephone companies were required by law to provide all necessary information and technical assistance to authorized law enforcement agents. The 1970 amendment was essentially the United States Department of Justice's response to an earlier incident, in which the Nevada Telephone Company had refused to assist intelligence officers perform a wiretap. Five years earlier, the company had been subjected to a $6 million lawsuit for assisting an FBI-sponsored wiretap operation (Carr 1998, Sects. 3.76–3.77; Lapidus 1974, p. 123).

That incident directs us, once again, to the complexity of the institutional forces in question. Despite assertions by critics, it would be a factual inaccuracy and political oversimplification to claim that the telephone companies have historically been prepared to consent to their role as aides to state-sponsored wiretapping operations. Sporadic in number, yet significant in symbolic value, refusals by companies to co-operate with the authorities have been recorded during the Cold War and even after the application of the 1970 amendment to Section 2518(4) (Carr 1998, Sect. 4.127; Dash et al. 1959). Such refusals do indeed make logical business sense: systematic assaults on the information security of telephone users can undermine the status of the network and, in turn, the revenue of the carrier (Brenton 1964, p. 162). It should not, therefore, be surprising that, ever since the onset of World War II, American telephone companies were known to be very stringent in the internal policing of their employees. The latter were continuously "exhorted on the theme of communications secrecy and made fully cognizant of state and federal laws dealing with the subject" (Brenton 1964, pp. 162–163; see also Lapidus 1974, p. 125).

Equally understandable is the persistent support given by the telephone companies to rules and regulations that enhanced the information security of their customers. As early as 1928, in *Olmstead v. United States*, a group of America's largest telephone companies acted in support of Olmstead by presenting an *amicus curiæ— friend of the court*, an official statement by a third party not directly related to the case, but whose interests can potentially be affected by the court's decision. The joint statement asserted that any form of wiretapping "violates the property rights of both persons using the telephone" (quoted in Murphy 1965, pp. 89–90). In 1963, William Powell, an influential New York Telephone Company executive, openly stated his complete disapproval of any form of invasion of the information security of telephone users, irrespective of the reasons behind it (Brenton 1964, p. 162). Four years later, on 18 May 1967, during the official hearings on privacy-invasion held by the United States Senate Subcommittee on Administrative Practice and Procedure, the vice president of AT&T declared that the Bell System supported a complete and general outlawing of all wiretapping, with the exception of interceptions performed on national security grounds:

> [p]rivacy of communications is a basic concept in our business. We believe the public has an inherent right to feel that they can use the telephone with confidence, just as they talk face to face. Any under-mining of this confidence would seriously impair the usefulness and value of telephone communications (quoted in Lapidus 1974, p. 126).

Historically, however, bound as it is by legal and institutional obligations toward the state, the telephone industry has been largely unable to continue to enjoy the state's preferential treatment, while protecting the information-security features of its communications networks from interference by state agencies. It has gradually learned to negotiate and protect its interests within the existing legal and institutional framework. For the most part, the telephone companies are thus "happy to be let off the hook" (Lapidus 1974, pp. 123–124). Their dealings with law enforcement agencies are mostly friendly and business-like (Dash et al. 1959, p. 246) and they assist the latter while striving to maintain a form of functional neutrality. In the words of an anonymous New York City telephone company official,

> [w]e try to minimize our role to the greatest extent. The less we know, the better, from the standpoint of law enforcement. The primary interest of the Bell System is the preservation of secrecy of communications to the maximum extent compatible with the public interest. Whether or not law enforcement officers should wiretap is a question for Congress to determine on a balancing of interests. Since Congress has passed Title III authorizing court-ordered eavesdropping, the Telephone Company accepts the law and assumes that it is constitutional (quoted in Lapidus 1974, pp. 123–124).

On balance, the historical attitude of the American telephone industry toward wiretapping has not favored the protection of the information security of users. There have indeed been instances where, upon discovery of illegal state-sponsored wiretapping performed under the assistance and supervision of company employees, the latter have immediately been dismissed (Dash et al. 1959, p. 246). Yet, the very necessity of maintaining public confidence in the telephone system has led the companies to conceal the existence and extent of legal and illegal communications interception, instead of exposing it. In the history of wiretapping, there has never been an instance when a British or American telephone company has reported an illegal telephone surveillance operation by government agencies, though many have been carried out. In fact, it has been asserted that the widely employed practice of law enforcement to use wiretapping as an investigative lead, as opposed to a method of collecting evidence for court, has not been the product of legal requirements. Rather, it is believed to derive from the constant demands by telephone companies to conceal the actual extent of wiretapping, in order to maintain public confidence in the telephone network (Dash et al. 1959, pp. 122–123, 146, 250).

Predictably, the dynamics of this situation have resulted in numerous instances in which telephone companies and their employees have repeatedly acted as agents of law enforcement and security apparatuses. Throughout the 1950s, telephone company staff in Philadelphia, Boston, San Francisco, Chicago, as well as in a number of cities in Louisiana, were consistently required to provide local and federal security officers with confidential subscriber information, temporary lines for wiretapping purposes, and even recordings of telephone conversations (Dash et al. 1959, pp. 123, 154, 165–166, 218–220, 246). Countless such instances were also reported during the 1960s. They included the case of C.V. Gris, a New York Telephone Company employee who, along with colleagues of his, ran for 2 years a sort of informal—and unquestionably illegal—wiretapping school for members of the police department, using telephone company facilities, equipment, and even funds (Spindel 1968, 112ff). In 1971 it was revealed that H.R. Hampton, a member of the executive board and director of government communications services of the Washington-based Chesapeake & Potomac Telephone Company, had perceived it as his "patriotic duty" to arrange tens of thousands of unauthorized communications interception operations for the FBI for more than 20 years prior to his retirement (Wise 1976, pp. 37–38n; Joy and Wright 1974, p. 257). Further enquiries into the matter revealed in 1975 that the vice squad of the Baltimore Police Department in Maryland

> had monitored telephone conversations with co-operation from the Chesapeake and Potomac [. . .] Telephone Company without proper legal authorization [...]. Staff of the phone company's Security Office provided members of local and state law enforcement agencies with non-published telephone listings upon oral request until October 10, 1973 (American Friends Service Committee 1979, p. 53).

Similar evidence emerged throughout the 1970s with regard to the underground connections between local telephone companies and FBI field offices in a number of American states (Ungar 1975, p. 448).

Another service by TSPs, which American law enforcement agencies came to expect during the Cold War, was alerting them whenever complaints were received by subscribers who suspected that their telephone lines had been wiretapped. Some scholars claim that this constituted standard practice with authorized or "national security" wiretaps (Brenton 1964, p. 163; see also Spindel 1968, p. 220; Dash et al. 1959, p. 123). Following his retirement, E. Belter, the chief supervisor of national security wiretaps for more than 20 years, explained in an interview that, when a telephone company received a complaint from a customer who insisted that their line was wiretapped, it would inform the agencies responsible. Then, "[w]hen it was cleared up [they] would say 'Okay, it can go back on'" (quoted in Wise 1976, pp. 53–54). Pennsylvania State University professor John Carroll asserts that, by the late 1960s, the situation had reached the stage where victims of wiretapping were advised by experts not to complain to the telephone company or the police, but to use information-security-enhancing instruments instead, such as voice scramblers (Carroll 1969, pp. 163–164).

Historically, the institutional ties between telephone companies and law enforcement agencies have also been sustained through the training of officers and the sharing of surveillance technology. Even before World War II, police and intelligence officers often enrolled for courses and seminars organized by local telephone companies for their own staff (Leonard 1938, p. 64; Dash et al. 1959, p. 72). In the postwar era it became customary for police and FBI communications-interception specialists to be recruited from the ranks of former telephone company technicians, repairmen and linemen with many years' experience on the job (Dash et al. 1959, pp. 50, 51, 223). Additionally, the wiretapping apparatuses used by police and intelligence officers were often manufactured, tested and sold to them by the telephone companies themselves (Dash et al. 1959, pp. 222–223; Leonard 1938, p. 64).

4.9 Discussion: A Debate That Isn't

One of the most frustrating elements in the history of domestic communications interception is that it is essentially not a dialogue: on the one hand, vocal critics of the practice insist on their claims. On the other hand, Home Office, or State and Justice Department officials usually resist being drawn into a discussion about the social, political or legal grounds of communications interception. Intelligence officials are legally forbidden from entering the debate, while law enforcement officers simply refer enquirers to legal experts.

Since the emergence of telephony, the few official responses to the public criticism of communications interception regimes in Britain and the United States centerd on attempts to rationalize it and to reconcile its somewhat unpopular nature with the more healthy political characteristics of parliamentary democracies. The logic of such attempts has usually been threefold. First, there is the argument that communications interception is inevitable. Every state does it. Throughout the twentieth century, telecommunications emerged as inseparable parts of everyday social processes, and as such have had to be policed. Second, it is proposed that law-abiding citizens, who exercise their constitutional rights and obligations while respecting those of others, have nothing to fear, because they have nothing to hide.

Finally, state-sponsored wiretapping is often viewed not as a right, but rather as an obligation—namely, the obligation of contemporary nation-states to ensure the security and welfare of their citizens and other residents.

On the other side of the debate, critics argue that there is something fundamentally wrong with models of state policing, or protection, which are not subject to popular control and accountability. In the United Kingdom and the United States, centralized institutions such as law enforcement and—especially—security agencies are not routinely subjected to direct electoral control. Furthermore, they have developed within a peculiar culture of bureaucratic exclusivity and institutional autonomy, in which notions of administrative transparency, political openness and moral self-evaluation have, for the most part, been regarded as weaknesses, rather than strengths. Relevant here are the words of former FBI Assistant Director William C. Sullivan, in his statement before the Church Committee in the United States Senate, in November 1975:

> during the ten years that I was on the US Intelligence Board [...] never once did I hear anybody, including myself, raise the question: 'is this course of action which we have agreed upon lawful, is it legal, is it ethical or moral?'. We never gave any thought to this realm of reasoning, because we were just naturally pragmatists. The one thing we were concerned about was this: will this course of action work, will it get us what we want, will we reach the objective we desire to reach? (quoted in Theoharis 1978, p. 94).

The idea that these institutions should be entrusted with the ability to invade the privacy of citizens, while, at the same time, operating under the considerable absence of popular and—in some instances—even executive forms of democratic control, is therefore seen as deeply problematic. One such critic, the British scholar Peter Gill, writes with reference to the United Kingdom:

> there is inadequate democratic control of state structures in the UK, particularly of the security intelligence agencies, and [...] this poses such a threat to democratic forms that it requires fundamental change (Gill 1994, p. 39).

Wiretapping, critics assert, was not inevitable during the Cold War. There were many states around the world that were known to have either outlawed it altogether or practiced it in conditions of strict legal and administrative control. In Belgium, a country deeply involved in the Cold War, state employees were heavily penalized for conducting wiretapping or for divulging, not only the content of a call, but "even metering information, such as the frequency, time and duration of calls, or the numbers dialled from a particular line" (Fitzgerald and Leopold 1987, p. 158). The penalties stipulated in the legislation—enacted in 1930—were substantially heavier for state employees than for ordinary citizens found to be engaging in wiretapping. During the 1950s, New Zealand enacted a similar legislation. In 1957, the country's Postmaster General stated in the House of Representatives that

> there is no telephone tapping in this country with one exception, and that is on the authority of the Prime Minister where the national security may be involved. In those circumstances and those circumstances only, will telephone tapping be tolerated. For the purpose of ordinary criminal administration there will be no telephone tapping (Dash et al. 1959, p. 300).

Clearly, critics exclaim, systematic wiretapping is not inevitable. In reality, it is a conscious political decision, which is largely shaped by the extent to which moral and political notions of respect of individual privacy inform the state's policy on information security.

The argument that law-abiding citizens have nothing to fear is simply dismissed by critics, who refer to the historical instances of political abuse of communications interception during the Cold War. The history of wiretapping during the twentieth century, they insist, clearly demonstrates that the practice was repeatedly used against law-abiding individuals and organizations. Scores of civil rights activists, pacifists, trade unionists, politicians, lawyers, actors, artists, feminists, students, homosexuals and environmentalists were repeatedly wiretapped by British and American security agencies during the Cold War, not for criminal, but for political purposes. Often, membership in a particular organization—as in the case of the Congress for Racial Equality in the United States—or even protesting against nuclear power—as in the case of the Campaign for Nuclear Disarmament in Britain—were deemed sufficient to justify lengthy wiretapping operations against individuals or groups. In the eyes of critics, the history of state-sponsored communications interception is the history of the undue invasion of privacy and the subversion of information-security principles.

Ultimately, it would be unwarranted to directly compare the peculiar social and political dynamics of the Cold War with those of today as an attempt to dismiss the practice of communications interception on political grounds. Yet the unfortunate precedence of that era of major political tension understandably haunts the contemporary debate on information security and cannot—or, indeed, should not—be ignored.

4.10 Summary: A Fine Line Between Discretion and Abuse

Soon after the emergence of telephony, its interconnection with everyday social negotiation was so drastic that neither the British, nor the American law enforcement and security communities were able to ignore its impact. On both sides of the Atlantic, the responsibilities of law enforcement and security organizations were sooner or later expanded to cover the policing of telephony. That expansion was indeed influenced by the peculiarities of the political landscape in each country. Thus, wiretapping acquired more openly aggressive characteristics in the United States—especially following the onset of the Cold War—while in the United Kingdom it assumed the discrete nature of many another governmental activity.

Yet in both nations, critics have not hesitated to virtually associate the history of wiretapping with the history of the undue invasion of privacy during the past 100 years. The debate is inevitably marked by political and ideological polarization, as well as by the absence of basic information. There is, however, one particular issue that might be drawn from it: the single fundamental factor that has historically facilitated the extensive use and, at times, misuse of wiretapping by law enforcement agencies in the United Kingdom and the United States has been the close formal and informal institutional relationship of those agencies with telephone companies and the telephone industry as a whole. This institutional interface is so critical that any future blueprint aiming toward the enhancement of the individual's right to information security in the digital environment is destined to be a disappointing failure unless it addresses the deep tacit link between the government and TSPs.

Chapter 5
The Techniques of Communications Interception

This chapter considers the technical evolution of communications interception. In addition to providing crucial technical background to the topic, it explains in detail the technological reasons that led to the emergence of RIPA and CALEA, as well as the structural changes that the wiretapping of digital communications posed for the actors involved in it. After a brief description of the technology of analog wiretapping, there will unfold an account of the barriers to communications interception posed by the digitization of telecommunications. Finally, changes in communications interception in the digital telecommunications environment—as facilitated by RIPA and CALEA—will be outlined and discussed.

5.1 The Micro-mechanics of Analog Communications Interception

From the emergence of telephony and until the onset of digitization, the basics of communications interception practice in the Plain Old Telephone System—known as POTS—environment remained essentially unchanged. In cases where legal parameters were followed, a communications interception operation would begin with a law enforcement or security officer securing legal authorization for intercepting communications data, or content. In the United Kingdom that was usually done under the Wireless Telegraphy Acts of 1949/1998, or the Interception of Communications Act of 1985. In the United States, such authorizations fell under Title III of the Omnibus Crime Control and Safe Streets Act of 1968, which was later amended by the Electronic Communications Privacy Act of 1986.

© Springer Nature Switzerland AG 2020
J. Fitsanakis, *Redesigning Wiretapping*, History of Information Security,
https://doi.org/10.1007/978-3-030-39919-1_5

5.2 DNR-Based Hardwired Taps

The authorized communications interception request was presented to a designated employee of a local exchange carrier, who was then mandated to provide administrative and technical assistance to the requesting officers. If the request involved simply the acquisition of *call data*—namely numbers dialled by the subject of an investigation during a prescribed period—then one of two things would usually occur: (a) if the request involved the suspect's past telephonic activity, then a printout of the suspect's subscriber records would be shared with the requesting officer; (b) if the request involved the suspect's future telephonic activity, then the requesting officer would use a particular kind of pen register device known as Dialled Number Recorder (DNR). Though antiquated, the term *pen register* is used to this day to describe the recovery and recording of the dialing information that addresses a telephone call to and from an intercept target. It stems from the manner in which the digits in a telephone number were recorded when telephones used pulse dialing technology. After 1963, pulse dialing was gradually phased out and replaced with dual-tone multi-frequency, which in the 1980s became widely known as touch-tone technology.

When attached to the target subscriber's telephone line, a DNR captured all electromagnetic impulses transmitted through the line, including impulses transmitted while in telephonic contact with another party (Freeh 2000). Phrased differently, while in operation, a DNR collected all of the contact numbers that were dialed by a target subscriber. It also printed lists of the contacted numbers accompanied by monographic symbols that indicated ringing, a busy signal, as well as the beginning time of the placement of a telephone call and the precise time that the called party answered (Yarbrough 1999). Some of the larger British and American law enforcement departments, as well as all security and intelligence agencies, had DNRs in their possession. In the last two decades of the twentieth century, DNRs were often used in parallel with computer software that was specifically designed to analyze and scrutinize telephone call patterns of targeted individuals or groups.

5.3 Cross-Connect Box Hardwired Taps

If the communications interception request involved accessing the actual content of communications, then a telecommunications employee disclosed to the requesting officers the particular cross-connect boxes—also known as Serving Area Interfaces—that mediated between the target subscriber's telephone apparatus and the telephone company's central switching stations, as well as with the binding post—that is, the identification number of the target subscriber's line. Once the requesting officers detected the precise cross-connect box in which they wished to install the wiretap, they notified the local exchange carrier employee, who then allowed them to lease one of the free lines available in the cross-connect box's 1800-pedestal—

named after the fact that a single cross-connect box usually provided service to 1800 telephone customers. With that information readily available, the requesting officers were able to install a wiretap without any further assistance from the telecommunications company. With the help of a pair of alligator clips, a hardwired connection was established between the target subscriber's line and the leased line. The latter channeled all telephonic activity on the target subscriber's line to a designated telephone located on the law enforcement or security agency's premises. Thus, in essence, if one considers a telephone connection between a subscriber and the network's operating center as a circuit consisting of a pair of copper wires forming a loop, an extra load—a component powered by a circuit—was connected to the circuit. That was done in a manner similar to connecting an additional telephone device onto a household's existing telephone line. The only difference was that, in a communications-interception operation, the additional telephone device was located at the law enforcement or security agency's premises. The additional telephone device used by the officers conducting the communications interception was usually an ordinary telephone appliance with its microphone disabled, so that it worked only as a listening device. It was usually connected to the audio surveillance equipment, such as a telephone recorder or even a typical voice recorder mediated through an automatic telephone-recording device. Additionally, a *slave* or *bypass* device was used to provide electrical isolation between the target line and the eavesdropping apparatus.

5.4 Loading Coil-Based Hardwired Taps

Another—though less popular—place to install a wiretap during a communications-interception operation was at the loading coil. Loading coils were induction devices placed by telephone carriers on those analog local loops that were longer than 5.5 km (3.4 miles) long. That was done in order to compensate for wire resistance and capacitance, and to boost the frequencies carrying voice information. Equipped with loading-coil and binding-post information provided by telephone company employees, law enforcement or security officers climbed the telephone pole and used alligator clips to connect a particular subscriber's line in the loading coil to a radio transmitter. The latter transmitted the required sound data to a recording receiver that was located within the law enforcement or security agency's premises. The radio transmitter was usually sheltered in a leather pouch known as a *disguise boot* and attached to the telephone pole or to the telephone lines.

Wiretaps installed on cross-connect boxes or on loading coils were known as *hardwired taps* because their implementation required physical access to a section of the telephone wire that sound signals travelled on. Law enforcement and security professionals were well accustomed to undertaking such communications-interception operations and telephone carriers were usually expected to provide only logistical support.

5.5   REMOBS Unit-Based Softwired Taps

A different wiretapping technique, known as *softwired tapping*, was developed soon after the emergence and use of the first mainframe computers for administering large-scale telecommunications networks. Softwired tapping has never been acknowledged in the United Kingdom, but in the United States it emerged during public hearings on telephonic surveillance following the Watergate scandal (see USCCGO 1974). This form of wiretapping required modification by the carrier's engineers to the software that was used to run a telephone system. In the 1990s, softwired taps became more widely known within the law enforcement and security communities as *translation taps*, Direct Access Testing Unit (DATU) taps, Electronic Switching System (ESS) taps or, more often, Remote Observance (REMOBS) taps. REMOBS were testing devices used by telephone engineers to run regular tests on local network loops, with each local loop consisting of around 12,000 customer lines served by a number of REMOBS units. Through these devices, authorized engineers maintained receive-only access to the content of any customer's line in the telephone system.

When served with a valid communications-interception warrant, the local exchange carrier's engineer supplied the requesting officer with the contact telephone number and access code of a REMOBS unit corresponding to the local loop that contained the target line. Once in possession of that number, the requesting officer used an agency telephone to connect to the REMOBS unit and keyed in the access code minus the last digit. After the code was accepted, the officer keyed in the target subscriber's telephone number. As soon as the last digit of the telephone number was pressed, the agency's telephone device became permanently connected to the target line and would not be disconnected until the last digit of the REMOBS unit's access code was dialed.

In the United Kingdom, prior to digitization, the vast majority of MI5 communications-interception operations are believed to have been conducted through REMOBS units, while in the United States REMOBS wiretaps rapidly replaced hardwired communications interception operations in the last two decades of the twentieth century. Financial reasoning was primarily responsible for this change in preference: REMOBS wiretaps were more economical than hardwired operations, since they did not require technical setting-up on behalf of law enforcement or security operatives. In fact, following the initial set up, REMOBS wiretaps were often conducted without any need for human attendance. Intercepted communications were automatically recorded on voice-activated magnetic tape or on digital recording devices and stored for later analysis.

5.6 Enter Digital Telephony

Starting in the early 1970s, telephone systems—including mainframe control and switching apparatuses, as well as transmission techniques—in the United States and the United Kingdom were subjected to a gradual process of digitization. The process has taken decades to carry out, and is not expected to be fully completed until after 2020 (Downes 2014; Hall 2018). In the meantime, digitization has been virtually transforming the nature of telecommunications networks by altering the technical paradigm of communications exchange. Prior to digitization, telecommunications systems were based on relatively simple strings of copper wires, which carried sound signals by translating them into electrical signals. Eventually these analog systems were enhanced by the addition of electronic amplifiers, in order to enhance voice signals travelling over long distances. They were further-improved with the addition of Frequency Division Multiplexing technology, which facilitated higher traffic on the analog networks. But these were trivial technical changes to the basic concept of network telephony that remained essentially unchanged since its emergence in the late nineteenth century.

By contrast, digital technology replaced the language into which sound signals are translated, by substituting electrical signals with binary signals. Thus, in the digital network environment, sound signals undergo a process of transformation from analog to binary code, using only the digits 0 and 1. This process is conducted by a series of electronic circuits, including *coders*. Coders have the ability to turn on—represented by the digit 1 in the binary system—and off—represented by the digit 0—extremely rapidly, thus translating all sampled sound signals into combinations of 1 and 0. Voice signals are converted—or *coded*—into binary format, which occurs at the local carrier's central office. They are then carried in digital format though a main trunk into the central office belonging to the local carrier serving the local loop, where the message's receiver is to be found. Once there, the signals are processed again, this time by electronic circuits known as *decoders*—commonly referred to, along with *coders*, as CODECs. These convert the message from binary code back into analog sound signals. Finally, the signals are carried through the local loop to the telephone handset of the receiver. The roots of this procedure, which is known as *digital encoding*, can be traced to competitive economics. Network systems based on binary logic are relatively simple to construct. They are reliable, have enormous capacity and—due to advances in the production of silicon circuits—are much cheaper to build and maintain than analog telephone networks.

5.7   Digitization as a Barrier to Communications Interception

One of the unintended side-effects of digital encoding is that it threatened to render traditional hardwired, softwired, and even DNR-based wiretaps virtually useless. Thus it threatened to have a potentially catastrophic effect on state-sponsored communications interception operations. The case of DNR-based wiretapping, arguably the most uncomplicated of communications interception practices, is indicative of the way in which the digitization of communications posed structural barriers to even the most basic analog communications interception functions. The automatic conversion of sound signals to binary signals meant that, once the telephone network's local loop lines were digitized, network processors would be unable to distinguish the telephone number used to facilitate a telephone call from the sound data transmitted during a call's duration. Phrased differently, digitization caused the disappearance of the operational line that has traditionally demarcated call-identifying information from the actual content of a telephone call. This major technical change made it impossible to distinguish between warrants that gave state agencies the right to access pen-register information—identifying data about the telephone subscribers, as well as the time and duration of a particular call—and warrants that allowed them access to the content of telephone calls.

   This problem also affected the realm of the Internet and Voice over Internet Protocol (VoIP) communications, where the emergence of—the now largely obsolete—Integrated Services Digital Network (ISDN) communication standards blurred the line that separates communications data from communications content. Consequently, DNR units were unable to collect the numbers of telephone calls made by target individuals or groups using fully digital telephone lines (Anonymous 1998a, 1999a; Dempsey 1999). In the relevant technical literature of the 1990s, it was generally recognized that the growing technological sophistication of digital telecommunications networks would only magnify this problem:

> [c]ommunication services are ostensibly for the conveyance of information between interfaces of a network but actually convey data streams, so there will be no indication of the information type contained in the data stream. The consequence is that deciphering the data stream will be problematic or at worst impossible (Parish 1999);

> [i]f [Asynchronous Transfer Mode, or ATM] becomes the enabling technology for the nation's next generation of multimedia networks, the [National Information Infrastructure], as some foresee, interception of electronic communication will become more difficult. Packet networks will require a substantially different approach to surveillance than used for today's digital telephony. Since the address is an integral part of the packet that contains the message data as well, it will be necessary to develop means to insert hooks into the packet header to identify the sender and the intended recipient (Blair et al. 1995).

Similar problems appeared in the other three forms of communications interception, both hardwired and softwired. The conversion of sound signals to binary signals in a digital network meant that, once the local loop was digitized, hardwired taps installed at cross-connect boxes would relay a series of unintelligible sounds—

similar to those one gets when listening into a facsimile, or fax, line—instead of a conversation between target subscribers. Theoretically wiretaps installed at cross-connect boxes were technically feasible if state agencies were in possession of Digital-to-Analog converters, such as those used by the carriers themselves, which could instantly translate binary signals back into sound signals. But even in that scenario only a small portion of the targeted communication would be accessible, due to a phenomenon known as *packet switching*. Packet switching is an ATM-based technique that fragments data streams into packets and transports them through different routes toward a common destination.

> A packet is a unit of data [representing a slice of speech lasting 30 milliseconds, which] leaves one site and traverses the network intact until it reaches the destination site [. . .]. The packet always contains some network address information to allow the network nodes to route the packet properly to the destination (Goralski and Kolon 2000).

Thus, the packetized transmission of a data stream on a digital network can be compared to the journey of a train through a railway network, in which all of the train's carriages are disconnected and allowed to make their way toward the point of destination through different routes. Shortly before arrival at their destination, the carriages are then assembled in the order in which they began their trip. All digital networks, such as the Internet, have evolved as packet-switched networks in order to facilitate constant user demand for faster and more efficient data communication. By the late 1980s, as large segments of the telephone industry were moving toward an Internet-like model of network communication, packet-switch solutions quickly became the norm in voice-data exchange.

Thus, by the mid-1990s, communications-interception models based on cross-connect box installations were proving increasingly insufficient in meeting the mandates of law enforcement and security agencies (Clayton et al. 2000; Dempsey 1999). According to some experts, the packetized nature of Internet communications rendered every known form of content interception almost practically impossible:

> [m]essages in transit on a backbone IP network are [. . .] difficult to intercept. There is no structure to the flow of data, with many thousands of 'calls' going on simultaneously. Indeed, the way the Internet is operated, with the possibility of alternative routes, means that the packets of data associated with any one transaction do not necessarily follow the same physical path between the end systems (Smith 1999).

In 2001, Rodney Small, an FCC employee who handled CALEA issues, warned that packet switching was expected to develop into one of the most corrosive elements in the communications interception debate between state agencies and telecommunications carriers:

> [w]e believe the packet issue is going to be around for a long time [. . . Industry has] decided it [i]s too expensive to [facilitate packet-switching-friendly communications interception] and they are not sure what the privacy implications are. They are getting cold feet, legally and financially. Meanwhile, these new technologies keep developing [. . .]. On the packet data [issue], there could be many petitions and it could be a big mess (quoted in Brown 2001).

Hardwired interception models based on loading-coil installations were also disappearing from use, since digital data transmissions did not require amplification from loading coils on lengthy wire routes. The same applied to softwired interception models facilitated through REMOBS units. By the late 1990s, most of the quality testing of digital telephony networks was facilitated through specially designed software that operated on centralized computer units, and thus REMOBS were rapidly disappearing from use (Yarbrough 1999).

5.8 Communications Interception and Custom Calling Services

In addition to the problems posed for communications interception by changes in the basic structure of wireline telephony, state agencies were concerned about a number of additional system features that were created or enhanced by digitization. These further-restricted the communications-interception needs of law enforcement and security agencies. Such features, which became collectively known as *custom calling services*, included some of digital telephony's most popular consumer preferences, such as call forwarding, speed dialing, three-way calling, automatic recall, and number portability. Essentially, the higher the ability of digital telephony subscribers to control the particular features and capabilities of their telephone service, the higher the level of their information security and the lower the ability of a government interceptor to gain access to their call data or content.

5.9 Enter Digital Wireless Telephony

The digitization of telecommunications also had a dramatic effect on the growth of mobile communications systems, such as cellular communications and Personal Communications Services. Such systems had been around since the 1970s and had always posed technical barriers to communications interception. Nevertheless, their extremely limited functionality and popularity prior to the late 1980s meant that mobile telephones were rarely targeted in state-sponsored communications interception operations. Digitization changed the nature of mobile communications systems in three main ways: (a) it reduced the cost of manufacturing mobile telephone handsets; (b) it reduced the cost of constructing and maintaining mobile communications networks; and (c) it helped increase the amount of data that could be transmitted through the network, which in turn decreased the costs of communicating through mobile systems. The above three changes were instrumental in facilitating the rapid increase in the use of second-generation cellular communications instruments in the United States and the United Kingdom throughout the 1990s.

By the end of the decade, cellular systems, regardless of whether they were analog or digital, functioned by dividing an area into small hexagonal cells. Each cell was sized between 2 and 35 km (3–50 miles) in diameter and was covered by a base station that consisted of a tower and a small building containing the radio equipment that was necessary for communication. This multiple cell structure allowed for extensive frequency reuse across a city or a town, so that millions of customers could use cellular telephones simultaneously. More significantly for communications-interception operators, multiple cell structures allowed mobile telephone users to *roam*—that is, to travel between an area's cells while using the same telephone handset. As a user moved toward the edge of a particular cell, the cell's base station sensed the diminishing signal of the user's telephone. At the same time, the base station of the cell that the user was moving toward sensed the increasingly strong signal of the user's telephone and gradually assumed the necessary reception and transmission functions. Often, the user roamed outside the coverage area of his or her contractual cellular provider, in which case prior industry agreements enabled another Mobile Switching Center to assume servicing the user's handset. In such cases, billing information was automatically exchanged between the original service provider's Home Location Register and the Visited Location Register.

The mobile nature of cellular telephony meant that it was impractical to place a hardwired tap at a particular cell base station, as there was no guarantee that the latter would facilitate all of a target's telephone call or calls. Indeed, by the late 1990s, a particular cellular telephone signal could be expected to pass through dozens of cell base stations and a number of different service providers before reaching its destination. Additionally, even if a target's cellular telephone remained stationary, by the time the voice signals reached the cell base station they would have already been converted to binary signals, since most cellular telephone devices were digital by that time. By 2001, virtually all second-generation cellular telephones in use contained built-in CODECs, allowing for digital-to-analog and analog-to-digital conversion to be performed at the actual handset. Thus, in the wireless environment, even the local loop was becoming digitized—something that was also gradually materializing in the wireline environment through ISDN standards. What that meant in practice for communications interception was that it often took law enforcement or security officers "weeks rather than minutes to decode the conversations" they intercepted (Lloyd 1993).

5.10 Battling for Standards

Throughout the 1990s, as increasing numbers of cellular telephony providers were switching from analog to digital, barriers to communications interception were multiplying by the week, and alternative models of authorized communications interception practices were being pursued through the close co-operation of law enforcement and security agencies and cellular carriers.

Both CALEA in the United States and RIPA in the United Kingdom aimed to counter the emergence of digital barriers to communications interception, by forcing the telecommunications carriers to provide law enforcement and security agencies with access to communications data and content regardless of its nature. None of the two Acts, however, specified the precise technical features of that accessibility. Rather, in both cases, the process of deliberating on the technical standards of digital communications interception came after the enactment of the legislation. Consequently, in the absence of concrete technical standards, large segments of the two Acts remained unenforced for years after they were enacted in to law. Predictably, following the enactment of RIPA and CALEA, the quest for acceptable technical standards emerged as the most contested issue in the digital communications-interception debate. Observers in both the United Kingdom and the United States remarked at the time that, although telecommunications carriers were aware that they were required to provide government agencies with access to digital communications data and content, they were simply in the dark about precisely how this access was going to be facilitated:

[o]f particular interest to suppliers of communications systems and networks is the technical means by which an interceptor will require to access a user's data stream. Will a system feature with very secure access be required? Are there service and interface descriptions? As currently drafted, the present proposals are too vague (Parish 1999);

nobody really knows how to do this. I mean, there isn't a blueprint, which has been worked out—at least not [. . .] one that is known about in the industry—of how this is supposed to happen (Bowden 2000, pp. 665–667; see also Interview 08 2000, 184ff);

[t]he [government] said [CALEA] required us to [. . .] put certain features into the technology and we said no, the law does not require those features. That was based upon a pushback on cost reimbursement grounds, as well as privacy grounds and [. . .], frankly, that issue is still open (Director, CALEA Compliance Unit, large United States TSP, Interview 01 2000, pp. 85–88);

it's just sort of this vicious cycle so that [. . .], even if a company wants to buy equipment [. . .] and become CALEA-capable or CALEA-compliant, they couldn't because there is nothing out there, because all the manufacturers are waiting to see what is going to happen [Government Relations Director, large national TSP association, Interview 18 2000, pp. 84–92].

In the United Kingdom, the technical features of RIPA remained largely unformed well into the 2000s, as the Home Office conducted "detailed technical consultation exercises with [TSPs] regarding the interception of new telecommunications technologies" (Home Office 2001b). Gradually, a Technical Advisory Board was created, consisting of six telecommunications industry representatives and six members of law enforcement and intelligence agencies. The Board was established under Section 13 of RIPA, which, among other things, considers the technical consequences of capability demands upon the telecommunications carriers (Home Office 2001a). Following initial uncertainty as to the degree of transparency that would characterize the Technical Advisory Board, it was agreed that its members should be security-cleared, that its meetings should be held in confidence, and that it should report directly to the Secretary of State (Home Office 2001a). It was therefore not until 2002 when observers without security clearances were able to form a basic picture of RIPA's technical features. It was immediately apparent, however, that the British government was instrumental in determining the specifics of RIPA's technical standards. The very fact that half the members of the proposed Technical Advisory Board were drawn from government agencies placed the British government in a position far superior to that of the United States government in relation to CALEA's technical standards. Specifically, Section 107 (a)(2) of CALEA contains what is often referred to as a "safe harbor" provision, which states that

> [a] telecommunications carrier shall be found to be in compliance with the assistance capability requirements under section 103, and a manufacturer of telecommunications transmission or switching equipment or a provider of telecommunications support services shall be found to be in compliance with section 106 if the carrier, manufacturer, or support service provider is in compliance with publicly available technical requirements or standards *adopted by an industry association or standard-setting organization*, or by the [FCC] under subsection (b), to meet the requirements of section 103 [Public Law 104–414, Section 107 (a)(2); emphasis added).

Phrased differently, the above quote means two things: (a) that, strictly speaking, there does not necessarily have to be a digital communications interception technical standard, as long as every carrier was willing and able to comply with CALEA in any way they saw fit; and (b) that if technical standards were to emerge, they had to do so out of an industry—not government—consensus. Hence, law enforcement and security agencies were not in a position to impose their own communications-interception standards on the industry. Notably, the safe harbor provision was not present in CALEA's early drafts; rather it was the outcome of intense and well-calculated industry lobbying efforts aiming to ensure that, as one lobbyist put it "law enforcement can't take us" (CALEA Lobbyist, large TSP association, Interview 16 2000, p. 239):

> according to the law [that] was enacted [law enforcement] cannot dictate to the switch manufacturers how to design the [CALEA-compliant] solution and they cannot dictate to [...] the carriers [...] how to provide delivery to them. They tried [...]. I mean, they [...] would love to. They even made 'suggestions' on how they would like to have it. But when [...] we make a decision, our decision is based on the best interest of [our company] (CALEA Administrator, large United States TSP, Interview 41 2001, pp. 29–34; see also Interview 18 2000, pp. 47–51; Anonymous 1997a; United States Court of Appeals 2000).

> Law enforcement [...] can't dictate how they have that [policing] presence [on the networks]. CALEA does [...] give the government the opportunity to [...] demand that our technology, as we change it, allows them to continue to conduct electronic surveillance; but they cannot tell us how to do that. It's a very [...] fine distinction, but it's very much present in the law (CALEA Compliance Attorney, large United States TSP, Interview 20 2001, pp. 47–49; 60–65).

In America, the safe harbor provision allowed the industry to generate and adopt an interim technical standard known as J-STD-025 (USTIA and USEIA 1997), which the law enforcement and security community described as "wilfully deficient in providing law enforcement the technical capabilities it needs" (FBI Special Agent, CALEA Implementation Section, Interview 19 2000, pp. 104–104) and rejected it as inadequate to protect public safety and national security. That confrontation resulted in a rupture in relations between the industry and the law enforcement and security communities, which led to several lawsuits and serious delays in the implementation of CALEA. During the deliberations, United States carriers repeatedly accused the FBI of delaying the construction of data collection systems, which meant that "even if a carrier were poised to install CALEA-compliant equipment, there would be no means for testing the equipment or even for law enforcement to receive any information once the equipment is installed" (Baker et al. 1998).

5.11   The Switch Solution

A basic principle of digital communications interception that guided RIPA and CALEA's technical standards was that, short of being able to install a listening bug on the suspect's telephone handset, the only place in the digital telecommunications system where a wiretap could be successfully installed was at the switching equipment. The latter is typically located at the local exchange carrier's central office, which is also known as the *telephone exchange* (Blair et al. 1995). Switches are complex computer instruments that replaced the role once held by human operators in the telephone system. In digital systems, all telephone numbers called within the vicinity of a particular local exchange carrier are channeled to the central switches, located inside the company offices. If the called party is located within the same vicinity as the calling party, the switching software is able to connect the two lines. If the two parties are not located within the same vicinity, then the switching software connects the caller with the central switches of the required vicinity. Without central switching equipment, the lines of all the subscribers in a telephone network would have to be interconnected, as was in fact the case in the very early days of telephony. The simplification of the network, which is facilitated by the existence of central switches, arguably renders the latter the most fundamental component—the structural core—of today's telephone networks. It is right at this core that communications interception needed to be facilitated in the digital environment:

> [law enforcement] realized that these new features and functions required a lot more flexibility [...] to target and that the only one who really knows what's going on is the switch (CALEA lobbyist, large TSP association, Interview 16 2000, pp. 23–29).

In the United States, where the quest for CALEA's technical standards was chronologically more advanced than United Kingdom's RIPA, the telecommunications industry proposed that switch-based communications interception should to be carried through a *delivery function* and a *collection function* system (USTIA and USEIA 2000, 17ff). Delivery function, namely an intercept port installed on a switch, would consist of two distinct types of channels: a *call content channel* and a *call data channel*, each intercepting a respective segment of target telephone calls. The intercepted data would be forwarded to the collection function, which would be controlled and maintained by the intercepting agency (Electronic Surveillance Task Force 1997). In a CALEA-supporting environment, the call collection function was less likely to be a DNR unit or a telephone recorder and more likely to be a high-speed personal computer operating on specially designed interception software. The same delivery and collection functions would be implemented in the cellular tele-communications environment, with call content and call data delivery channels applied to the Mobile Switching Centers of cellular carriers (USTIA and USEIA 2000, 83ff).

That proposal to use switches as the physical locations of digital wiretaps was more significant than it may initially appear, because it implied three fundamental changes to the technology and practice of communications interception. First, for switches that were installed prior to the implementation of CALEA and RIPA— namely the vast majority of switches—it would be impossible to establish a hard-wired tapping connection to the central switch without causing a breakdown to the hardware's programmable functions. Instead, the order for a wiretap had to be given through the computer software that monitored the performance of the switch. Since the monitoring computers were located within the carrier's central offices, it was designated carrier employees who would have to install the wiretap pertaining to a communications-interception request. Therefore, in essence, the traditional role of the carrier in communications interception operations would be revolutionized, as the latter would assume not only logistical but also operational responsibilities under the proposed regime.

Second, as a direct result of the carrier's augmented role in communications interception, the role of law enforcement and security personnel in communications interception operations could be heavily compromised. Under the proposed regime, the latter would have had virtually no input in, or control over, the technical installation of a wiretap. Instead, a leased line provided by the local exchange carrier would provide them with the required telephone call data or content. Interestingly, in the United Kingdom, this structural change in communications interception operations had been observed since the mid-1990s. Writing for the political review *The New Statesman and Society* in 1994, intelligence correspondent Patrick Fitzgerald noted that

> [t]he introduction of digital switching in the telephone system has changed the method of telephone-tapping by British Telecom which it provides on an agency basis to MI5, Customs, Special Branch, the National Criminal Intelligence Service, and police forces. Under the old 'Tinkerbell' system, tapping circuits had to be physically connected to targeted phone lines. Now 'taps' are simply set up using the software programs within the exchange switching system, moreover this is done remotely without ever going near the telephone exchange building. The 'requesting agencies' increasingly have intercepted calls relayed directly to their headquarters where they can analyze and transcribe telephone calls (Fitzgerald 1994, p. 30).

Third, prior to CALEA and RIPA, digital central switches were not designed with interception in mind (Interview 04 2000, pp. 272–275; Interview 17 2000, pp. 12–22). Thus the industry's future standard blueprint for central switching mechanisms would have to be altered to reflect RIPA and CALEA's recommendations. Namely RIPA and CALEA-compliant switches would have to be designed with an additional port that would allow for the parallel feed of intercepted data. Considering the functional centrality of switches in the digital telecommunications network, it was obvious that wiretap-friendly technological alterations in the central switches would directly subvert the most basic principles of information security.

5.12   United States: Intercepting the Internet with DragonWare and DCS 1000

Even following the conception and introduction of CALEA in the United States, any form of comprehensive communications interception of Internet traffic was considered by the industry as an exceptionally challenging technical task. Nevertheless, on 6 April 2000, a lawyer testifying before a United States House of Representatives subcommittee alerted observers to the existence of "a proprietary software programme with the not very reassuring name of Carnivore', which was used by law enforcement and intelligence organizations to intercept data circulating on the Internet. The lawyer went on to say that

> [a]lthough the government acknowledged that Carnivore would be capable of capturing more than the information authorized under the order, it would be programmed to obtain only information from the target subscriber's account, and would be configured not to intercept the content of any communication. It was acknowledged that Carnivore would enable remote access to the [Internet service provider's (ISP)] network and would be under the exclusive control of government agents (Corn-Revere 2000).

In October 2000, following that initial revelation, and under legal pressure by Freedom of Information Act applications filed by civil liberties groups, the FBI released a number of documents pertaining to the program. The documents were heavily redacted, with more than half of the 750 pages blacked out and hundreds more withheld (Lemos 2000). According to the documents, Carnivore was conceived in February 1997 by engineers working for the United States law enforcement and intelligence communities, and was first used in October of that year under the name Omnivore (Meeks 2000). This complex technical device was then replaced by a user-friendlier version, which was named Carnivore. In February 2001, following internal criticism of the name's aggressive connotations, it was renamed once again, this time to DCS 1000, with DCS standing for Digital Collection System.

Initial information provided by the FBI said that DCS 1000 was an Internet interception device used by the law enforcement and intelligence community in cases where ISPs lacked the technical ability to facilitate authorized communications-interception requests. As explained earlier, the packetization of digital communications constituted one of the primary barriers to intercepting digital messages. Consequently, as the number of communications-interception requests issued to ISPs by governmental agencies increased over the years, so did the number of cases where the service providers were unable to decode packetized data. That phenomenon triggered the creation and use of DCS 1000 (Kerr 2000). Thus, according to Michael Warren, Senior Project Manager of the FBI's CALEA Implementation Section, the technical novelty of DCS 1000 was not that it was able to intercept packetized data, but rather that it was able to decode the information contained in them and distinguish between data referring to call-identifying information and to call content (Cisneros 2000).

Initially, outside observers were under the impression that Carnivore was an email surveillance technology that was able to intercept information pertaining either to the sender and receiver addresses of an email message, or to the actual content of the email, depending on the level of authorization permitted by the interception warrant. A few months later, however, once further technical documents were released under the Freedom of Information Act, it became clear that Carnivore was a mode of function of a larger Internet surveillance system known as DragonWare Suite. DragonWare consisted of three software programs: (a) Carnivore, which collected email data and content; (b) "Coolminer', which collected Internet pen register data; and (c) "Packeteer', which used data collected by both Carnivore and Coolminer to reconstruct web sites exactly as they were when a target Internet surfer accessed them (Meeks 2000).

The FBI, which was the only United States government agency that admitted to possessing DragonWare technology, refused to reveal its source code and technical specifications. But it acknowledged that Carnivore—DragonWare's primary interface with the Internet—was produced from commercially available monitoring tools, such as EtherPeek, or the Ethereal freeware, which was used by ISP network administrators for network oversight and management. Virtually no information was disclosed about the technical features of the hardware that connected to the network and collected the intercepted data. Nevertheless, it can be safely inferred that the device was small enough to be portable, and that it was enclosed in a box so as to be protected from physical attack or mistreatment. The box almost certainly contained switch-like components, including a hardware authentication device used to control access to the box, and a network isolation device, used to prevent the box from transmitting to the intercept target.

The FBI indicated that Carnivore could be installed on ISP *backbones*—large transmission lines that carry data gathered from smaller lines that interconnect with them (Kerr 2000). However, doing so would only increase the amount of data that had to be filtered by the device in order for the precise packets of targeted information to be identified. Thus, the ideal location to place the device would have been near the ISP servers. The latter are programs that serve the files that form web pages for users; their role is thus to facilitate the communications of an Internet user. To make Carnivore even more effective, it would have to be placed right after the ISP's router, which is a software-led device in a computer that determines the next network point to which a data packet should be forwarded toward its destination. Once Carnivore was installed, it scanned all incoming and outgoing data for messages associated with the target of a communications interception operation (Srinivasan

2000). It then "surgically" (Kerr 2000) intercepted and forwarded copies of these messages to the intercepting officers. Considering the above, DragonWare could be perceived in schematic terms as a bridge—a connector between a local area network and another local area network that uses the same protocol—which collected, copied and forwarded all traffic from a subscriber's line to a central collection point over leased forwarding facilities.

5.13 United Kingdom: Active, Semi-Active and Passive Interception on the Internet

At the onset of digitization in the United Kingdom, the debate on Internet communications interception largely concentrated on the evaluation of three potential models of interception, none of which appeared to replicate DragonWare. The three models were proposed by an independent marketing group commissioned by the government to conduct a technical and financial study of RIPA (Smith Group 2000). The report termed the three models as (a) active; (b) semi-active; and (c) passive interception, in accordance with the degree of involvement that they required from of ISPs.

Under the active interception scheme, which was applicable only to email surveillance, a software-based wiretap would be configured by authorized ISP personnel at the company's email server. Targeted incoming and outgoing mails would then be automatically copied and forwarded by the server to a "government host"—possibly the National Technical Assistance Center (NTAC), which the British government established in 2002—which could be as rudimentary as a government email address (Smith Group 2000, p. 38; see also Ward 2001; Home Office 2001c, d). This suggestion was strongly preferred by the British government, as it was technically efficient and appeared to be causing little displeasure among the industry:

> [i]f the interception aim is modest, just to read the target's email, then the task may be simple. Even with permanent connections and certainly with dial-up accounts, the ISP will be running a store-and-forward system for email. Making a copy of the email as it passes through the ISP system is straightforward—you just need a configuration file that says this is to be done. Such a file can be trivially created by any administrator when served with a suitable warrant (Brown et al. 2000, p. 7).

Under the semi-active interception scheme, a data collection unit would probably be housed at GCHQ in Cheltenham (Clayton et al. 2000) and would be permanently hardwired into the network. Additionally, ISP routers would be configured in such a way that a number of pre-selected IP addresses, reserved for users who are under authorized surveillance, would be routed through the collection unit. Hence, once a communications interception operation against a particular subscriber would be initiated, the ISP would be notified and the subscriber would be assigned one of the pre-selected IP addresses. In this way, the pre-determined IP addresses are guaranteed to be routed via specific routers at an initial interception point, at which the ISP interception device sits. It will be ensured that subscribers are not able to determine whether or not their traffic is being forced to pass this point (Smith Group 2000, p. 39).

Some ISPs preferred this solution to the passive interception scheme. This was because, even though reconfiguring the network's routers was a complex task, in the long term the amounts of data that would have to be filtered by the collection unit would be relatively small. Thus, the network's overall speed and efficiency would not be threatened (Clayton et al. 2000).

Similarly to the semi-active interception scheme, the passive interception scheme also required a data collection unit to be permanently hardwired into the network. The difference was that, under the passive scheme, the unit would continuously mediate all Internet traffic passing through the network. While allowing data sent or received by non-target users to pass through without being tampered with, it would automatically copy and forward to a predetermined location all data sent from target subscribers. Under that scheme, government agencies would be in charge of providing, installing and maintaining all hardware and software components of the interception system and would also implement communications-interception operations against users without having to inform ISPs of these users' IPs or names (Smith Group 2000, p. 37).

5.14   Towards a Permanent Interception Presence

A particularly significant variation between the proposed Internet interception schemes under CALEA and RIPA concerns the uneven degree of permanence that characterized them. In the United States, DragonWare and its interception point, Carnivore, were not permanent features of the telecommunications network, but were instead deployed and installed to facilitate ephemeral communications interception operations. By contrast, both the semi-active and the passive interception models proposed in the United Kingdom under RIPA had their basis on a conception of interception units as integral components of public telecommunication networks. Independent observers criticized such plans, insisting that.

> [i]n practical terms, this requires the development of black boxes, installed at each ISP, that will send traffic data to the [National] Technical Assistance Center, located within the Security Services (Brown et al. 2000, p. 5).

The British government's plans were often compared to the implementation of Russian Internet interception system SORM, (translates to System for Operational Investigative Activities) which began to be implemented in Russia in 1995 and continues to operate today as SORM-3. Under this system, all of Russia's ISPs are legally bound to install an interception unit that permanently links them to the Russian Federal Security Service. One critic of the British government's proposals said that the similarity between RIPA and SORM was "striking. RIPA will mandate black boxes at all ISPs. [We are] following the path set by Russia when it comes to interception almost exactly" (Brown et al. 2000, p. 33).

5.15 Summary

There was a time when state-sponsored communications interception was technically plain and simple. That time was brought to an abrupt end by the onset of digitization. The latter was seen as injecting unmatched dynamism to the newly denationalized market. At the same time, however, it severely compromised the communications interception mandates of law enforcement and security agencies. This compromise, which for a while threatened to drive the interception arms of these agencies out of business, eventually led to the emergence of RIPA and CALEA. Under these proposed regimes, the structure of communications interception would change dramatically, marginalizing the role of government agencies and centralizing the role of the carriers in the implementation, oversight and maintenance of wiretaps. In addition, certain elements in digital communications-interception models would require the redesign of structural components of the digital telecommunications network, so as to render it interception-friendly. This applied not only to the telephone network, but also to the Internet, where new government-sponsored technologies were seen to gradually overcome the network's design barriers against communications interception.

Chapter 6
The Views of the Negotiators

This chapter presents the findings of interviews conducted with some of the main actors involved in the digital interception debate in the 1990s and early 2000s in the United Kingdom and the United States. Material collected from other primary and secondary sources is also exhibited in an attempt to outline vital elements in the digital communications interception debate that unfolded in the United Kingdom and the United States as digitization was emerging.

6.1 Interception of Communications as a Social Tool

There is a lengthy historical tendency within the British and American law enforcement establishments to regard communications interception as a fundamental pillar of investigative competence. Undoubtedly, intelligence organizations have traditionally been much more familiar than police agencies with the culture of communications interception as an information-collecting tool. This is because of the sheer volume of information that is surreptitiously acquired, processed and analyzed by intelligence agencies, as opposed to the more public and visible communications role of policing bodies. Nevertheless, the increasing reliance of models of policing on communications and information management, of which communications interception is part, should not be underestimated. In their groundbreaking ethnographic work on the communications practices of Canadian police, Richard Ericson and Kevin Haggerty describe the role of policing in a risk society as one of brokering to large administrative institutions and organizations that operate on the basis of risk knowledge. Thus, the optimization of communications-interception practices and techniques by increasingly organized and bureaucratized law enforcement bodies is a direct consequence of the perceived need by those bodies—and by society at large—to respond to the external demands for risk knowledge, defined by the authors as the societal process of accessing and communicating vital information about danger and risk (Ericson and Haggerty 1997, Chap. 1). The authors go so far as

© Springer Nature Switzerland AG 2020
J. Fitsanakis, *Redesigning Wiretapping*, History of Information Security,
https://doi.org/10.1007/978-3-030-39919-1_6

to assert that the particular features of communications systems, including surveillance systems, actually govern institutional relations within law enforcement bodies as well as their efficacy and success in relation to their perceived aims (Ericson and Haggerty 1997, p. 4).

In the United Kingdom few—if any—public references are ever made to the extent of usefulness of communications interception for law enforcement. In the United States, however, federal agencies repeatedly stress the operational value of communications interception. In 1995, the then-Director of the FBI Louis Freeh remarked that communications interception was

> the single most effective investigative technique used by law enforcement to combat illegal drugs, terrorism, violent crime, espionage and organized crime (Freeh 1995).

Law enforcement interviewees in the United States wholeheartedly confirmed Freeh's statements. In responding to the hypothetical question "what would happen if the right of law enforcement agencies to engage in communications interception was revoked by Congress?', one federal law enforcement interviewee responded:

> I think I'll be safe in saying that the voice of law enforcement would [. . .] cry out if there was any revocation of that authority to do intercepts [. . . W]e would have a hard time recovering. Law [. . .] enforcement has depended so much on surveillance—and I'm talking about telephonic intercepts—as a means by which to gain enough evidence to convict people; it has been an incredible tool for us and without it we would be lost when it came to complex investigations (FBI Special Agent, CALEA Implementation Section, Interview 19 2000, pp. 258–262, 275–276).

Another interviewee, a former undercover police officer with many years' experience in operations involving communications interception, said that the operational value of communications interception is so crucial that he never even thought of working without it:

> I've never even thought how hard it'd be to [. . .] infiltrate a group of criminal or espionage agents without the ability to intercept their private communications [. . .]. It's the 'A' of our intelligence-gathering alphabet [. . .] (Police Officer with experience in undercover and wiretapping operations, Interview 14 2001, pp. 33–36).

Similar answers by British law enforcement interviewees further support this interdependency between criminal or intelligence investigations and communications interception:

> I do not believe law enforcement could carry out its role without the ability of interception. In terms of when to intercept, that is debatable. But the actual facility to be able to intercept I think you have to have it [. . .] The law enforcement agencies have got to have some way to be able to get into those types of organizations and interception is just one of those (Public Affairs Officer, MI5, Interview 33 2000, pp. 6–14).

This interdependency is not seen as inevitable; rather, it is imposed upon law enforcement agencies by lack of financial resources and the increasing emergence of advanced telecommunications systems in modern societies. A police officer with lengthy experience in communications interception operations, admitted that you haven't always had electronic interception; law enforcement was still able to get results before wiretapping. But back then you didn't have telephone communication systems [in] everyday use either (Police Officer with experience in undercover and wiretapping operations, Interview 14 2001, pp. 33–36).

A regulation affairs officer with a large British TSP added that communications interception is that the illicit use of telecommunications generates the need for networks to be policed:

I think that communications interception has got quite a long history. It's been related to voice telephony until recently, so [...] the need is driven by the use people make of communications in society (Regulatory Affairs Officer, large British TSP, Interview 11 2000, pp. 9–11).

Historically, this view of communications interception as a required element of policing in advanced societies has rarely been questioned by the carriers and providers of telecommunications on either side of the Atlantic. In the United States, the telecommunication industry's collective representation bodies are often eager to stress their historical commitment to allowing law enforcement access to their communications networks. In 1997, during the height of the CALEA debate, Roy Neel, President and Chief Executive Officer of the United States Telecom Association, said that "the telecommunications industry has always cooperated, and is committed to continuing to cooperate, with lawful and authorized law enforcement electronic surveillance activity" (Anonymous 1997a). A year later, he added that America's

telephone companies have worked side by side with law enforcement for many years to provide them with surveillance technology and access to caller information. We take this responsibility very seriously and are committed to assisting law enforcement agencies with crime prevention and detection measures (Miceli 1998).

A similar outlook permeated the attitude toward communications policing of collective industry organizations in the United Kingdom in the 1990s (Parish 1999; Emery 1999; Sutter 2001). Indeed, interviews with leading industry representatives in both the United States (Interview 20 2001, pp. 112–114; Interview 21 2001, pp. 20–21) and the United Kingdom showed that there was little variance of opinion among providers and carriers in regards to law enforcement's right to maintain a structured presence in communications networks. An interesting distinction can be noted in that United Kingdom industry officials appeared to be motivated to accommodate law enforcement needs by virtue of legal requirement, whereas their United States colleagues stressed corporate etiquette and civic duty as chief sources of their motivation:

I think traditionally the telecoms industry have always recognized that they're in a business which is required to provide this sort of service to the authorities in the areas of commerce [Officer, MI5, Interview 12 2000, pp. 6–14; see also Interview 04 2000, pp. 35–36];

we have a role as corporate citizens to be co-operative and we're just as interested in helping law enforcement catch the crooks that use [...] our phone systems [CALEA Compliance Attorney, large United States TSP, Interview 20 2001, pp. 121–124];

I don't think there is a company that we represent that does not agree that they have a social obligation to supply law enforcement with information on criminal activity [Technical Director, national association representing over 500 smaller TSPs, Interview 17 2000, pp. 53–55].

Broadly speaking, however, the interviews showed that, both in the United Kingdom and the United States, moral and civic duty considerations were more prominent in the public rhetoric—if not the decision-making process—of carriers than official documents tended to show. In the United Kingdom, this was recognized by the Information Commissioner, who stated that, while communications data handovers remain voluntary, carriers are placed "under considerable moral pressure to co-operate with police" (Sutter 2001). Symbolic terminology, including phrases such as "common good" (Interview 04 2000, p. 303), "civic duty" (Interview 06 2000, p. 153), "good corporate citizen" (Interview 11 2000, p. 63; Interview 12 2000, p. 535), "social responsibility" (Interview 17 2000, p. 36) and even "national security" (Interview 16 2000, p. 117) were repeatedly mentioned during interviews as important ideological guidelines shaping the carriers' relationship with law enforcement:

we get requests to run through our data to give them call data information [...], like, for example, when we had the [...] Olympic bombings in Atlanta in 1996. We put forth a lot of effort. We ran call records from all the payphones around the Centennial Park, where the bombing occurred, to assist them in developing leads [...]. We do a lot of things, [be]cause we are good corporate citizens. We try to be [CALEA Compliance Attorney, large United States TSP, Interview 20 2001, pp. 263–268].

Interviews showed that, in essence, United Kingdom and United States carriers did not consider the practice of communications interception for law enforcement purposes to be fundamentally different from other public-order provisions by governments, such as physical patrol policing (Interview 04 2000, pp. 15–17), traffic policing (Interview 07 2000, pp. 70–74), gun control (Interview 08 2000, pp. 8–13), health and safety regulation (Interview 10 2000, pp. 550–558) and emergency service provision (Interview 12 2000, pp. 63–67). "In other words', said a law enforcement liaison of a large British ISP association,

in a sort of parallel universe, policemen need police cars to go chasing bank robbers and chasing people exceeding the speed limit at the motorway; and nobody [...] has ever suggested to me that the police don't need police cars to drive around in. They should be allowed to do that. They should be allowed to break the speed limit if [...] it seems like a good idea at the time and jump red traffic lights and [all] that (Regulation Officer, large British ISP association, Interview 10 2000, pp. 30–35).

Nevertheless, British carrier officials appeared more willing than their United States counterparts to distinguish significant political implications in the potential abuse of communications interception, which inevitably sets the practice apart from public order provisions by the contemporary nation state:

we're moving from an era where only the bad guys—in theory—or political subversives or whatever, but very few people [...] had anything intercepted [...], to one where everybody is intercepted on all their communications. Because they're going to do it; they're going to put this black box in here, it's going to be crammed in the switch and it'll be intercepting everything [Chief Executive Officer, large British ISP, Interview 08 2000, pp. 429–434; see also 191–196; Interview 10 2000, pp. 42–48, 633–636; Interview 04 2000, pp. 18–21].

Similar opinions appeared to inform some of the less established ISPs in the United Kingdom, which were typically encountered in responses by the industry to government consultation documents:

> ISPs would like to be able to reassure their users that no automated set up of interception will be requested. Although more convenient to ISPs and absolving their employees from jeopardy on 'tipping off', automated interception would raise a significant 'Big Brother' image of the Internet (Anonymous 1999a).

Such awareness of the extraordinary social and political significance of communications interception use and abuse was often substantial and appeared in the form of moral dismissal of overly aggressive communications interception practices and demands:

> I don't think we morally should [change our network technology to accommodate communications interception]. I mean, if you want my opinion, I think intrusive surveillance, in [. . .] the [RIPA] terms, [. . .] is a bad thing for society and that [. . .] I would find that morally difficult to cope with, if we were doing that [Security Officer, large British TSP, Interview 07 2000, pp. 212–215].

Broadly speaking, therefore, ISP representatives say themselves and their companies as obligated to assist the state's policing functions as part of their civic duty. At the same time, however, the seemed acutely cognizant of the extent to which communications interception could cause a 'chilling effect' in a democratic society, and objected to such a scenario of telecommunications policing.

6.2   The Technological Promise of RIPA and CALEA

Undoubtedly, the driving force behind the conception and implementation of RIPA and CALEA was the realization that the process of digitization of telecommunications had largely taken place without consideration of the communications-interception requirements of law enforcement and security agencies. Thus, it was the evolution of the technical features of telecommunications that gave rise to an extensive restructuring of communications interception in the digital environment, as reflected in the RIPA and CALEA legislations. Carriers on both sides of the Atlantic appeared fully aware of the technological motivations behind RIPA and CALEA, whose "promise" (Interview 16 2000, pp. 212–216) was to return large segments of the digitized telecommunications realm to governmental supervision and control:

> I don't think it's any secret that we already extensively assist the police services and [. . .] other people in this area. What this new legislation does, really, is [. . .] to put some new vocabulary and some new vernacular around something, which is happening already [Security Officer, large British TSP, Interview 07 2000, pp. 1–4].

CALEA didn't change the substance of law. In other words, [. . .] what law enforcement has to do to convince a judge—what is [. . .] required for the judge to issue a court order—did not change [. . .]. All CALEA did was [. . .] it clarified [. . .] what carriers had to do to assist law enforcement in the conduct of electronic surveillance. And part of that clarification was that, as your technology changes in your network [. . .], technology has to continue to accommodate electronic surveillance. That's all CALEA did. It's a simple concept [CALEA Compliance Attorney, large United States TSP, Interview 20 2001, pp. 281–289].

In other words, according to the TSPs, CALEA and RIPA did introduce a new mode of surveillance on the network. Rather, they ensured that the state's right to access the pattern and content of telecommunications activity continued into the twenty-first century.

6.3   The Ability to Intercept in the Digital Environment

There was less certainty among carriers about whether the digital telecommunications environment was more susceptible to enhanced and structurally seamless communications interception. Interviews reflected the technical literature in this respect (see for instance Blacharski 1998 and Rozenblit 2000), by generating diverse views on the subject:

nowadays [the network] is infinitely more transparent in some ways than it previously was [. . . T]here's a huge amount of information stored, which the old analogue systems would not have [. . .] captured. [S]o [in the] normal course of business, everything you can conceivably think of gets almost accidentally logged. So [. . .] I can't see us having to change our network technologies to accommodate the RIPA [Technical Specialist, large British TSP, Interview 25 2001, pp. 204–211; see also Anonymous 1997a and Kampschoer 2000];

But recording subscriber activity as part of day-to-day business and giving outside actors—such as government agencies—access to them were two different things. In a letter sent in 2000 to the British secretary of state, a telecommunications industry executive remarked that

the trend is not towards networks that are easier to intercept; rather the reverse is true. The techniques needed for interception—finding where the packets are, specially identifying them, then duplicating them to special places—are not compatible with high-speed low-cost networks. Efficient and secure networks are inherently hard to intercept. Even if ISPs are currently able to intercept at the edges of their networks, those edges are blurring and moving ever outwards towards the customers themselves [Lansman 2000].

In the United Kingdom, many carriers—mostly ISPs—appeared to be sharing the view of a number of civil-liberties activists who claimed that "it is not [technically] possible to arrange anything like a comprehensive intercept capability" (Bowden 2000, pp. 667–670) in the digital telecommunications environment:

[one ISP] has had situations with police officers demanding actual web pages [accessed by] their customers [. . .], which would have been absolutely impossible for their size of customer base and also they wanted [. . .] complete details of emails or contents of emails

going back to [. . .] 10 years or whatever—for as long as they've been operating. [That was]
just technically unfeasible. And, you know, the police officer thought 'I can do this, because
I have a warrant' [Fraud Control and Revenue Assurance Manager of large United Kingdom
wireless communications company, Interview 04 2000, pp. 181–189];

with regard to carrying out interceptions in the 'middle' of larger networks, there is a
unanimous view amongst industry experts [. . .] that such interception verges on the impossible and will certainly bear significant costs [Lansman 2000; see also Parish 1999].

In the United States, where CALEA implementation began in 1995, carrier
representatives appeared more prepared to overcome or bypass some of the more
complex technological barriers to state-sponsored communications interception,
providing financial resources from public funds were made available to them:

I think you need to tie up [technical and financial issues] together. Anything [. . .] is
technically feasible. It's at what price is it technically feasible and who's going to pay for
it? So, the simple answer is [that] anything could be done technically. It's how much do you
want to pay for it and where do you get your money back? [Policy Lobbyist, national
association of rural TSPs, Interview 38 2000, pp. 160–161, 171–173];

there are a few things that law enforcement wanted that [. . .] would affect carrier networks
that [. . .] we pushed back for technical reasons. But you pour enough money on us and we
can probably solve the technical problems [CALEA Compliance Attorney, large United
States TSP, Interview 20 2001, pp. 361–363].

Predictably, this view was supported by United States law enforcement interests,
who refused to consider the technological complexity of digital systems as grounds
for rejecting CALEA-compliant solutions (see Morris 1998):

believe me, [CALEA] is not a technical nightmare for [carriers]. This is all well within their
capabilities. That [. . .] has never been a big problem. Other than the cost of doing, the
technical ability to do it . . . all the packet-protector whinnies are saying 'hey, we can do this;
we can do a lot more than that'. Yeah, this is easy [FBI Special Agent, CALEA Implementation Unit, Interview 19 2000, pp. 610–614]

One British industry representative stated that "in some instances we'd possibly
be able to get [law enforcement] more [information]—what they haven't even
thought of" [Interview 04 2000, pp. 178–179]. Parliamentary records show that
followers of this viewpoint were also to be found within the British government.
Former Minister of State Charles Clarke likely reflected a general policy principle of
his Office when he stated in Parliament that he did "not accept the suggestion that the
age of interception is over or that work in that area may be futile. That', the Minister
of State continued,

would be a philosophy of despair. I understand why people take that view—they see the
Internet as an all-embracing vehicle that cannot be controlled by any national legislature. We
believe strongly that the power of this form of communication and the nature of global crime
in so many important areas, such as drugs, pedophiles and so on, is such that we must do
what we can to control it. The powers in the Bill will allow us to have an impact on that. To
say that we cannot do anything would be to accept a counsel of despair and we cannot accept
that [HCSCF 2000].

Admittedly, Clarke viewed the advancement of state-sponsored digital communications interception as a primarily political, rather than strictly technical, executive

policy decision. Prior to RIPA's implementation, the absence of specific technical information prevented detailed evaluations of the legislation's technical feasibility. However, some relevant information was included in a leaked report produced by the Internet Service Providers Association (ISPA) through a secret forum that brought its representatives in contact with the Association of Chief Police Officers (ACPO) in November of 1997 (Akdeniz et al. 1999). The report showed that, according to network technicians, there were at that time very few digital barriers to digital communications interception that could not be overcome by technical means. The report stated that, although substantial technical improvements were required to increase the reliability of Internet-based communications interception, most of law enforcement's requirements—such as tracking Internet Message Access Protocol (IMAP) or Post Office Protocol Version 3 (POP3) messages through packet snooping, identifying IP addresses and monitoring customers' web site visitations— could be met, if the active co-operation of the relevant ISP and the installment of specialist monitoring equipment, was financially compensated from public funds. According to Yaman Akdeniz, whose organization, Cyber Rights & Cyber Liberties United Kingdom, publicized the leaked document, the report

> showed [...] that anything is possible—technically—to do. [ISPs] can monitor everything, [...] not only emails, but all sorts of wave activity, Inter-Application Communication, etc. I didn't know at that time that it was technically possible to [...] do all sorts of [...] activity [Akdeniz 2000, pp. 349–352].

Broadly speaking, however, interviewers noted that, while a number of technical barriers to interception could be overcome in the short term, the introduction of further, digitally enhanced service features in an increasingly competitive telecom-munications market would certainly debilitate even the most sophisticated communications-interception models:

> I think [the government ha]s set its agenda out and it's got a big wish list. And some of it will be possible, some of it will be possible but only for a very short period of time and other things will be 'yeah, that's great to have' [Fraud Specialist, large telephone service provider, Interview 24 2001, pp. 265–267].

Technical details aside, it appears clear in interviews that carriers were aware that telecommunications digitization had been anathema to state-sponsored communica-tions interception. This had not necessarily occurred due to intrinsic and inevitable security features in the digital environment, but rather due to the absence of law enforcement interests from technical debates on telecommunications digitization. Industry interviewees on both sides of the Atlantic seemed well-informed about the particular features of the digital telecommunications environment that denied law enforcement access to communications content, and in some cases even call data. Features such as VoIP, multi-path, packetized voice communications, cellular and pay-as-you-go telephony, broadband and narrowcasting technologies, as well as call-conferencing, call-forwarding, speed dialing, voice dialing, automatic redial and radio-pager technologies were mentioned in several interviews (Interview 04 2000, pp. 267–272; Interview 08 2000, pp. 410–413; Interview 10 2000, pp. 22–24; Interview 11 2000, pp. 284–286, 537–538; Interview 16 2000, pp. 119–125; Interview 17 2000, pp. 203–205; Interview 20 2001, pp. 131–134) and reflect official responses from carriers to government consultation documents (Anonymous 1999a, e). In regards to teleconferencing as a debilitating feature in communications interception, an interviewee stated that he was among the first American federal law enforcement officers to notice something was amiss, as early as the 1980s:

> we became concerned about this when we observed these long periods of inactivity on the telephone when we knew [the targeted individual] was talking to someone on the telephone line. We could either observe it or we had [...] a microphone in there or an informant. We knew that he was using the telephone but for some reason our equipment was not working and not picking this up. I guess it first came to my notice when we've had prisoners [...] in a local penitentiary there making calls to an associate only to be conferenced by that associate to a third party. And the associate would then drop out of the conversation or [...] would hang up his phone [...]. Of course, the wiretap would be on the associate outside of the prison. And we would be left high and dry with no content of the conversation, while all of the time knowing that he's talking to a drug associate or something and that the conversation was being supported by these associate's facilities and services at the phone company. We would [...] certainly get the billing record of it, but we had no idea what was said (FBI Special Agent and Telecommunications Industry Liaison, CALEA Implementation Section, Interview 40 2000, pp. 30–38).

It was therefore through such incidents, which happened while at work, that government agencies such as the FBI began to confirm the growing gap between their technical ability to intercept and the rapidly evolving digital features of the telecommunications networks. It is also worth noting that in some cases the criminals appear to have become cognizant of these changes faster than law enforcement.

There also appeared among carriers a certain apprehension of a number of emerging digital features, which were expected to pose drastic *cul-de-sacs* for communications interception, and whose emergence had been technically confirmed but had yet to be addressed by communications-interception legislation:

> there is a need to change all this, but then the question arises: well it's all going to change again [...] within the next three or four years; how do you write legislation which has got to be future-proof, but is not just a blank check for the Secretary of State [...] to do what he feels like at the time? [Regulatory Affairs Officer, large British TSP, Interview 11 2000, pp. 290–294];

> [t]here is no question that [...] the technological explosion that we're going through right now is going to continue even at a faster pace [...]. When we get fiber to the home, then what are you going to do? There's no [...] switch point; there's no place to put [a tap]. [...] I think there are more questions that are going to be asked and there are very few solutions that are going to come forward [Policy Lobbyist, national association of rural TSPs, Interview 38 2000, pp. 208–215];

> with ISDN coming up, with fiber to the home, I mean, it's kind of tucked into that. You can't go inside a person's home and run some beads out the back door [...] into a recording device [Director, National Services Planning, large TSP association, Interview 37 2000, pp. 32–34].

> I think the interesting thing about legislation and regulation is [...] that technology is constantly outpacing regulations and legislation and you can see with CALEA that [...] they never foresaw the [...] problems they're having with [information technology]

(Government Relations Director, large national TSP association, Interview 18 2000, pp. 175–181).

Law enforcement interviewees acknowledged the relative incapacitation of traditional technical models of communications interception caused by the dynamics of digital telecommunications evolution. Interviewees in the United Kingdom confirmed rare public acknowledgements by government and law enforcement representatives (see Todd 1999, Veigas 1999, House of Commons 2000, HCTISC 1999a) of serious setbacks to state-sponsored communications interception:

I'm not going to say what our blind spots are, because I [...] wouldn't want the wrong people find[ing] out how to evade supervision. [B]ut I will say that in the [19]90s we suffered the most astonishing blows to our interception capabilities in living memory— certainly since I joined the Force [Communications Intelligence Officer, MI5, Interview 02 2000, pp. 110–114].

British Telecom has your number now, you decide you don't want British Telecom any more, so you move it. And you go to Cable & Wireless and so on and so forth. Which means that [...], from a law enforcement point of view, it's a nightmare—you know? Who actually does own that number? And it's even worse when it comes to mobiles [Officer, MI5, Interview 12 2000, pp. 494–498].

In the United States, law enforcement and security services were more willing than their British counterparts to share such frustrations in public (see Morris 1998, n56; Freeh 1997; United States Court of Appeals 2000; Blair et al. 1995; Kallstrom 1997). In 1994, after receiving the analyzed data of a nationwide survey on communications-interception complications encountered by law enforcement in the course of investigations (see Boucher and Edwards 1994, Sect. 14–15; United States Court of Appeals 2000), the leadership of the FBI said it was concerned with "losing wiretap capability entirely" (Dempsey and Oram 1998). FBI Director Louis Freeh stated at the time that the Bureau was "missing a part of the playing field, but our position is that we don't want to miss the whole playing field" in regards to communications interception. Freeh was speaking in light of a—classified at that time—1992 report by the FBI's Advanced Telephony Unit, which warned that, in the absence of the implementation of a CALEA-type program, the Bureau would lose access to between 40 and 100% of communications targeted for interception by 1995 (Diffie and Landau 1998, p. 183, n274).

Even the NSA, arguably the technical and engineering vanguard of the United States communications intelligence complex, admitted in 2001 that its operational capabilities were under threat by the digital telecommunications evolution:

[w]e're behind the curve in keeping up with the global telecommunications revolution [...]. Our adversary communications are now based upon the developmental cycle of a global industry that is literally moving at the speed of light [...]: cell phones, encryption, fiber-optic communications, digital communications [NSA Director Michael Hayden, quoted in Verton 2001].

United States law enforcement interviewees described some of the more contested digital features of new telecommunications networks, such as VoIP, as a "huge can of worms" (Interview 19 2000, pp. 526–527) and admitted to suffering from lack of technological foresight even following the implementation of CALEA:

> [when we were drafting CALEA], never did we think that telecommunications over a period of time—as we know it—would be supported by an ISP [FBI Special Agent, CALEA Implementation Section, Interview 19 2000, pp. 531–532];

> [a]fter CALEA came about, there was a huge sigh of relief [. . .]. We thought we had it covered. And then, about a year later, our consultants started calling us with [. . .] news that some ISPs were moving into the voice telephony market [Police Officer with experience in undercover and wiretapping operations, Interview 14 2001, pp. 89–92].

The impression one gets from these statements is one of a never-ending cycle, in which the incessant pace of technological change, coupled with the business-like speed and creative chaos of deregulation posed new problems for law enforcement and intelligence agencies every few months. This was especially true of the technology of telecommunications, which seemed to have no apparent end point. Instead, the constant pace of competition between hardware manufacturers and service providers was taking place in the near-complete absence of security and law enforcement requirements. The latter came in later, in the form of government-imposed legislation.

6.4 Law Enforcement's Knowledge Gap

Particularly prominent in interviews with British industry representatives were manifestly negative attitudes over what many perceived to be an absence from law enforcement operative standards of even basic technical knowledge about telecommunications:

> law enforcement people [usually] turn up with a bit of paper asking for something in gobbledygook, because they don't understand, or turn up and, as a lot of people in the industry say, their first words are 'I don't know anything about the Internet, but can you do this or that or the other for me?' [. . .]. I [have] been taken along to one of [law enforcement's] operation centers that deals with this kind of thing and was just amazed [by] how out of date everything was, how it was all—you know—how unsophisticated it was. And [. . .] this is what you're likely to find, rather than all of us being spied on by secret satellites that can read our number plates from five miles in the air. I think that most of the time you'll find that police [officers] don't even understand that a '.com' email address [...] is something to do with the United States [Regulation Officer, large British ISP association, Interview 10 2000, pp. 328–332, 347–355; see also 344–347; Interview 04 2000, pp. 191–193].

According to interviewees, the underlying reasons for this phenomenon are primarily cultural, and have much to do with the casual pace of change that is characteristic of bureaucratic governmental institutions. In other words, law enforcement bodies have been given little incentive by government to change their investigative and organizational routines:

> the police are [. . .] almost virtually seen as not moving on in their thinking about crime from the [19]60s or [. . .19]70s. Sometimes you feel as if [. . .] they're only really interested in that

type of [low-tech] crime. When it comes to more technical crime, because it's more to solve or more difficult, they can't possibly [...] solve it themselves [Fraud Control and Revenue Assurance Manager, large British wireless communications service provider, Interview 04 2000, pp. 58–64];

the police [...] still—by the way—refer to us [as the] GPO [(General Post Office). ... T]he [y] still phone us up and say 'is that the GPO?'. And that is [laughs] and that was before I joined the company—it was the GPO. [M]ost policemen think that the Post Office deals with the post and that the GPO deals with the phone. And so, they are so far removed and they don't understand what's happened [...]. At the working level in the police forces they have no clue as to what's happened. They just assume that everything is as it used to be [Technical Specialist, large British TSP, Interview 25 2001, pp. 237–246].

This belief is generally shared by British civil liberties lobbyists (Bowden 2000, pp. 380–381, 426–429) and appears to confirm similar opinions that permeate written responses by carriers to the United Kingdom government's consultation document on the Interception of Communications Act (IOCA), a largely unsuccessful pre-RIPA attempt to tackle early communications digitization that was introduced by the government in 1985 (Sutter 2001).

At the moment, the big problem ISPs have with the police is their stupid questions. After a while, it gets expensive and unproductive. It's a problem. It's always a problem, and it's a very serious problem [ISPA member Tim Snape quoted in Delio 2001; see also Anonymous 1999a, b].

Accordingly, numerous United Kingdom ISPs requested that, prior to fully implementing RIPA, the British government proceeded to train law enforcement officers sufficiently on communications interception-relevant technical issues:

ISPs would expect to deal with technically competent [law enforcement] staff, who are aware of the relevant Internet protocols and their properties. We are not in business to provide free training for law enforcement. Our experience in the past is that this is an important practical issue and the government should consider making the use of centralized expert assistance, from [law enforcement] or elsewhere, mandatory when a local police force wishes to implement an interception order [Hall et al. 1999; see also Anonymous 1999b, 2000b; Perry 1999; Weatherall 1999].

Such attitudes did not surface in interviews with United States industry officials, nor did they appear in the American literature on the subject during the time. It should not necessarily be assumed, however, that American law enforcement agencies enjoyed a higher degree of approval from carriers when it came to their technical expertise.

6.5   Shifting the Communications Interception Paradigm

In the late 1990s, some British ISPs insisted that it was indeed the lack of technical understanding by law enforcement of digital telecommunications networks, as well as their refusal to engage in sophisticated technical training, that ultimately shaped the British government's RIPA methodology (Anonymous 1999a). Hence, instead of opting for a communications interception model based on network backbone interception by law enforcement agencies, the United Kingdom's RIPA strategy appeared to be to "spread the pain out and [. . .] require every ISP to maintain some sort of interception capability" (Bowden 2000, pp. 650–655).

This interception model, which, as explained earlier, characterizes the basis of both RIPA and CALEA, implies that, for the first time in the history of communications interception, 100% of a wiretapping operation has to be performed by the carrier (Yarbrough 1999, Sect. 6; Kerr 2000; Charney and Alexander 1996). Both law enforcement and carrier interviewees appeared to recognize this crucial paradigm shift in communications interception:

> [i]t used to be [that] we'd provide the logistics—the information on the subscriber's line—and [. . .] that was it. We'd never hear from them again, unless something went wrong or whatever [. . .]. Now [the carriers] will have to do everything—they'll have to tap the line on the screen and either route [the intercepted data] to a separate location or even record it themselves. And [then] they'll have to untap [the line] when they're told to do so and—you know—everything shifts to the carrier [Former British Telecom external maintenance engineer, Engineering Division, Interview 03 2001, pp. 481–488; see also Interview 04 2000, pp. 244–247; Interview 07 2000, pp. 259–261; Interview 12 2000, pp. 559–560; Interview 16 2000, pp. 186–193; Interview 21 2001, pp. 122–123];

> for the [. . .] drug dealer, the pedophile and that type of person, even down to the racist, to the football hooligan—that sort of activity—yes, we do need telecom industry input, because we haven't got the resources otherwise [. . .]. And, you know, without telecoms' input [we] would just turn out the light [Public Affairs Officer, MI5, Interview 33 2000, pp. 565–569];

> for years, we would be out climbing the telephone poles. We would be out in the cross-connect boxes [. . .] and we would be armed with information they provided us over a phone call, saying, 'hey, Joseph's phone lines are here at this intersection. [They are] at this particular box on the ground there and if you go to cable number' so on and on, this number . . . this number 'and you clip onto them you'll hear Joseph talking'. I'd go up there [. . .] or send my guys out there, we'd clip on a clipper and then the carrier is none the wiser. I mean they know that it probably is out there somewhere but they're not doing it. And it was pretty much . . . you'd wash your hands clean of it. Now—wow—it's completely different in several respects: one, they have to do all the work. I call them up [and] say, 'look, I need this intercept. You've got 'x' number of days to get this thing up'. And they're struggling to do the work in their switch—with some keystrokes. And they're held directly liable to the courts. We're not doing anything. Say they're not doing it correctly and hook it on to the wrong person. It's their problem, yeah. So the shift of labour is [. . .] going to be dramatic [FBI Special Agent and Telecommunications Industry Liaison, CALEA Implementation Section, Interview 40 2000, pp. 652–669].

The above statements describe a new interception model that recognizes the lack of technical knowledge by law enforcement and intelligence services, and simply transfers all technical functions related to communications interception to the service providers.

6.6   TSPs as Guarantors of Communications-Interception Legality

During the CALEA debate in the United States, the FBI went to great pains to draw a firm distinction between the act of technically implementing a communications interception request and the act of qualitatively evaluating its necessity. According to the FBI, the fact that the carriers' role would be more ingrained in the mechanics of the communications interception process under CALEA, should not entail that carriers could assume the responsibility of assessing the justifiability of communications interception requests:

> [it should not be] suggest[ed] to carrier personnel that they are supposed to test the legal process [of communications interception operations] against some 'look up table' of statutes, which are often somewhat complex, and then substitute their review for that of a judge when a carrier is presented with a facially valid court order. Carriers are the implementers, not the enforcers, of lawful intercept orders, authorizations, and certifications under the electronic surveillance laws in this regard [Morris 1998].

There were numerous early signs, however, that carriers were being drawn into a more active and engaging legalistic role in communications interception operations, which was a direct outcome of their upgraded functional role in the implementation of communications-interception orders. In the United Kingdom, British Telecom went so far as to request assurances for the personal safety of any of its employees "involved in security-sensitive cases, such as those involving terrorists" under RIPA (Stewart 1999). Additionally, interviewees both in the United Kingdom and the United States suggested that carriers were already being drawn into court battles as witnesses against communications interception targets prosecuted by the authorities:

> I have even reached the point of giving [the police] my personal information—my date of birth, name, etc.—because the [intercepted] data I gave them are going to be used as an exhibit in a court case and I need to be there and confirm it [Fraud Liaison Officer, large British TSP, Interview 09 2000, pp. 235–238];

> yes [we co-operate with law enforcement], quite regularly—most notably [with] the Computer Crime Unit. We've had quite a lot of involvement [in] helping them with gathering evidence—that kind of thing [Fraud Specialist, large British telephone service provider, Interview 24 2001, pp. 129–130];

> if we're out there installing the taps then [we are] sometimes required to testify in court that we did do so and that, as far as we're concerned, we got the right individual [or] group of individuals [...], that we heard them speak the words recorded and so on [...]. Now [the carriers] are gathering the data on our behalf, so, yes, their people will be called [...] to appear in court and confirm their role in the operation [Sergeant, large United States police department, Interview 15 2000, pp. 37–43].

It is perhaps not surprising that, faced with such a high degree of engagement in communications interception operations under RIPA and CALEA, British and American telephone carriers were actively querying the justification of communications interception requests:

I want to hear some background. I want to know the seriousness of the crime so I [can] decide how much of my resources I'm going to make available. This is manpower, right? If you tell me [. . .] 'this guy is up for murder, we believe he [. . .] did the whole thing off the Internet', [then] all right, we'll just sit here and see if we can do it now. Right. That's different [. . .]. But, occasionally, guys come here with a warrant [. . .] and it should be: 'have you got a warrant?' and [if] you've filled it out wrong [then] this is a violation of my client's human rights. Or you may [. . .] have filled the form in right but you don't have enough justification [. . .]. Yeah, you filled in the form, you've got all the boxes right, but you've asked for this [much] whereas the crime is only that [small]. [So], my client is not going anywhere. He has got rights [and] you haven't shown me enough justification [Chief Executive Officer, large British ISP, Interview 08 2000, pp. 309–311, 340–349];

in the past, there has always been a break—i.e. the industry has done one thing, be it that it has to have a warrant—but really at the end of the day we've no idea why that warrant was served or why that warrant was invoked. Where[as] here we are starting to get a much clearer picture as to why we're being asked. And if we're not happy, the procedures are there for us to go back and say 'I'm sorry, I don't know what your warrant says, but [. . .] unless you can convince me that I'm not breaking this rule, this rule, this rule, then I 'ain't executing' (Regulation Officer, large British TSP, Interview 31 2001, pp. 381–384; see also Interview 12 2000, pp. 276–282);

by virtue of the services and features we provide, in the future, electronic surveillance is going to have to be conducted out of the switches as opposed to on the [. . .] lines out of the switch. [It]'s going to be controlled by us, so I think that does compress the potential areas of compromise. And so I think that makes it more secure and more private [CALEA Compliance Attorney, large United States TSP, Interview 20 2001, pp. 293–296].

In a 2000 presentation before a large group of law enforcement liaison officers of United Kingdom wireless and wireline carriers, the United Kingdom telecommunication industry's designated Single Point of Contact under RIPA characteristically described the new, upgraded communications interception role of carriers as guarantors of communications interception legality:

[t]he nature of [a RIPA communications interception] enquiry is in detail. What material benefits [are] the law enforcement seeking to gain from this request? [W]hat these new procedures are doing, is starting to make them sit back and identify why they are asking. If it happens that the procedure has not been followed and if you have suspicions along those lines, then it is up to you to bring them to our attention, and the necessary investigations can take place and, if necessary, the wrongs will be made right [Anonymous 2000b].

To a large extent, this novel attitude stemmed from the carriers' need to safeguard their legal status by ensuring that, while fulfilling law enforcement requirements under RIPA and CALEA, they were also able to meet the information-security demands of their users. This applied particularly to the United Kingdom, where civil liberties organizations (Akdeniz 2000, pp. 239–244), telephone carriers (Interview 08 2000, pp. 225–230) and even law enforcement themselves (Gaspar 2000, Sect. 7.1.1) acknowledged at the time that certain elements of RIPA clashed with the Data Protection Act 1998 and the European Convention on Human Rights (ECHR) protocols.

As far as service providers were concerned, that clash could potentially compromise revenue. Carriers on both sides of the Atlantic generally appeared to view the increased burdens placed on them by communications interception legislation as efforts by government and law enforcement to make them, not only technically, but also legally responsible for communications interception in the digital telecommunications environment, in terms of privacy laws:

> it's a pendulum and we're sitting in the middle. If it all goes wrong [. . .]—say there [are] two big buckets of water and you say 'I was wrong'—we know who's going to get wet [Chief Operating Officer, large British ISP, Interview 26 2001, pp. 337–339];

> [a]fter three years of futile negotiations, the industry is caught in the middle and faces massive sanctions despite our good faith efforts to implement the [CALEA] statute [President and Chief Executive Officer of USTA quoted in Miceli 1998].

In short, CALEA and RIPA required an unprecedented degree of involvement by TSPs in the facilitation of communications interception. The government claimed that this involvement would be strictly technical. However, the carriers feared—and were in fact beginning to see—that their involvement in the facilitation of communications interception blurred the lines between law enforcement and telecommunications service provision.

6.7   TSPs as New Trenches of Communications Security and Control

This new positioning of carriers at the operational center of models of communications-interception practices in the digital telecommunications environment was significant: it was expected that the new regime would treat carriers as crucial brokers of influence and control by groups or institutions wishing to either increase their communications interception capabilities or augment the security of their communications. The United Kingdom-based Alliance for Electronic Business (AEB) correctly remarked that, under a communications interception model whereby all carriers and service providers are required to facilitate state-sponsored communications interception, the risk of abuse could be substantial:

> [a]ny system capable of intercepting and understanding communications could become a target for those working outside the law [Anonymous 1999e].

AEB's concerns were not unsubstantiated; in 1992, the Head of British Telecom's Security Department admitted that his Department spent "a third of [its] time and resources on looking at [British Telecom's own] employees who are allegedly dishonest" (Bleep 1992). Reflecting AEB's concerns, some British carrier interviewees pointed to the considerable potential of telecommunications employees being in a position to systematically undermine communications interception investigations and even questioned the trustworthiness of their own staff:

> **Q:** Do you think that there are individuals operating within telecommunications companies on behalf of criminal syndicates?

> **A:** Of course, of course, primarily [. . .]. Especially engineers.

Q: Are you positive about that?

A: Yes, indeed. In the past there have been cases of carriers where insiders were assisting criminal elements [. . .]. Do you realize how easy it is to be working for a [carrier] company and [. . .] lending a hand to criminals in return for a commission? I caught a guy here once; I certainly did [Chief, Fraud Department, large British TSP, Interview 28 2001, pp. 156–165, 366–368];

it's not beyond the boundaries of imagination that you've got a Mr. Big. Our local Mr. Big is a couple of blocks away. He drives around in a great big car, has loads of money, but the only income he's supposed to get is £400 a month. Now, if our local Mr. Big was kind of concerned about what he was doing and was using us as an ISP, it wouldn't take him long to work out who the man that handles all the logs and things is, who's the only person that's authorized to turn the key on the black [interception] box and download the data. So it's not [. . .] going to be hard for him to make sure that that guy [. . .] gouges all the tapes and he puts them in the cupboard; or gets the tape and replaces it with a little blank one and changes the label; or forgets to do the backup on his messages—you know? [Chief Operating Officer, large British ISP, Interview 26 2001, pp. 386–395; see also Lansman 2000].

These concerns were echoed in the United States. In the early 2000s, American industry officials alerted officials to the potential of internal, carrier-based undermining of state-sponsored communications interception operations. In 2001, Grant Seiffert, Vice President of External Affairs and Global Policy, United States Telecommunications Industry Association noted that,

[i]n a packet [communications] world, somebody has to open the packet to look for the information the FBI is seeking. Is the FBI going to do it [. . .]? Who is going to be looking over everyone's shoulders when we open up this information? (quoted in Brown 2001).

But American telecommunications and law enforcement officials appeared to be more worried about the prospects of large criminal cartels or adversary intelligence organizations actually purchasing locally-based TSPs in order to shelter their communications from state-sponsored wiretapping:

in effect—I've made this comment to the Bureau that I'm sure they were well aware of— since the Telecoms Act of 1996 has come about, who is to say that drug dealers or the mob cannot go out and create their own—or have not gone out and created their own—competitive local exchange carrier? They can do that themselves. I am sure they have thought of that. It would take a lot of financial resources to do that, but you're talking about folks that have strong financial backing. They launder a lot of stuff [CALEA Project Manager, large United States TSP, Interview 21 2001, pp. 146–151].

Indeed, an FBI Special Agent interviewed for this project revealed that there had been at least one case where a local exchange carrier was targeted and purchased by organized criminal interests wishing to communicate in a communications interception-free environment:

[w]e've actually seen the Mafia—the Italian Mafia—purchase a telephone company in New Jersey—the whole telephone company. Now, how are we going to tell that? You can't just knock on the door and say 'we've got this court order and we were wondering if you could help us, but don't tell anybody' [FBI Special Agent, CALEA Implementation Section, Interview 19 2000, pp. 716–719].

In the United Kingdom, service provider ICL raised this very issue with the government:

> [w]e can expect smaller and smaller organizations, even individuals, to operate their own e-mail systems rather than relying on ISPs. It is likely that conspirators will operate their own services too, thereby either hiding key messages from surveillance or warning the targets of the surveillance. Any interception is going to have to rely fundamentally on the trustworthiness of the ISP and we expect this assumption to become more doubtful over time (Emery 1999).

Notably, the FBI remarked in Congressional hearings that it expected foreign ownership of US-based telecommunications carriers to have a negative effect on communications interception operations, because foreign-owned carriers would not necessarily be expected to share the United States law enforcement and security community's sense of safeguarding national security and public safety. In 2000, Larry Parkinson, FBI General Counsel, testified that a

> significant area of concern [for the FBI] is the security of United States intercept and data acquisition activity [. . .]. Without adequate safeguards, the damage to an [interception] investigation would be done almost from the moment the United States serves process on the foreign affiliated carrier (STTCPUSHRCC 2000; see also Anonymous 1997b).

Ultimately, it should not be presumed that concerned British and American state institutions, including law enforcement, were content to place increasing communications-interception operational burdens on privately owned telecommunications carriers. That paradigm shift, which could conservatively be characterized as drastic, signified an unprecedented loss of control by state institutions of the micro-mechanics of most domestic communications interception. FBI sources have expressed regret over that, by admitting that

> [b]ecause law enforcement no longer has direct access to the intercept point, it can no longer determine for itself on a continuous basis whether the delivery channel or circuit is operating properly [Yarbrough 1999; see also Morris 1998; Blair et al. 1995].

In interviews with law enforcement officials in both the United States and the United Kingdom, it clearly emerged that they were in considerable discomfort over the loss of the direct management of communications interception operations:

> changes in technology [. . .] mean that we have, in essence, lost control over a tool [that is] crucial to us; we can still get it done—hopefully—but we're not the ones doing it [. . .]; we are effectively sub-contracting [. . .]. Is this [a] change for the better? I'd be hard-pressed to say 'yes' [Detective, large United States police department, Interview 36 2000, pp. 67–71; see also Interview 19 2000, pp. 681, 703];

> I mean, you don't really know these [carrier] companies—you have no idea who they are, who they work for. Security will be compromised, no question about it [. . .]. It is only a matter of time [Intelligence Officer, National Criminal Intelligence Service, Interview 22 2001, pp. 404–406].

There is clearly a sense in which neither state agencies nor the telecommunications industry are particularly excited about the marriage of necessity that is represented by CALEA and RIPA. However, given the state of technology, as well as the needs of the state, the new placement of carriers at the operational center of dicital communications interception is seen as inevitable—if potentially problematic.

6.8 The TSP/Law Enforcement Interface Under RIPA and CALEA

The close historical ties between law enforcement agencies and telecommunications carriers or service providers in the United States and the United Kingdom have been discussed in earlier chapters. That was indeed confirmed in numerous instances by law enforcement and industry interviewees on both sides of the Atlantic (Interview 02 2000, pp. 93–96; Interview 09 2000, pp. 263–265; Interview 10 2000, pp. 332–333; Interview 17 2000, pp. 142–144; Interview 18 2000, pp. 61–62; Interview 19 2000, pp. 447–449; Interview 21 2001, pp. 22–25; see also Anonymous 1999c).

Both in the United Kingdom and, especially, the United States, carriers have traditionally endeavored to retain an "arm's length" (Interview 21 2001, p. 39) distance from law enforcement institutions, and there was skepticism about whether that traditional distance would be retained in the RIPA and CALEA environments. Both carriers and law enforcement representatives appeared to recognize that their institutions were, in a sense, "compelled by legislation to work together" (Interview 06 2000, pp. 163–166):

> [t]he combined efforts and collaboration of the industry and the law enforcement agencies will likely be required on a continual basis for the foreseeable future as the nation's communication infrastructure undergoes a nearly complete metamorphosis (Blair et al. 1995; see also Veigas 2000).

Interviewees confirmed that, in the pre-RIPA and -CALEA environments, the relationship between law enforcement and carriers was not a particularly defined process (Interview 07 2000, pp. 186–187), and had generally been maintained through a variety of informal connections, including personnel exchange between law enforcement criminal intelligence and telecommunications security departments (Interview 09 2000, pp. 343–345; Interview 14 2001, pp. 213–216) and even through large-scale projects, such as British law enforcement's Joint Co-operative Community Approach (Interview 07 2000, pp. 190–193). An MI5 interviewee explained that,

> [i]f you want an investigator it will cost you the Earth to take a man off the street and train him. If you want an investigator into telecoms, there is no such animal. So what you do is you say 'well, what's easiest?'. The easiest is to bring them on board and bring them up to speed in terms of the telecoms side, so let's take an investigator who has cut his teeth in law enforcement, be it customs, be it police, be it intelligence, and has those skills—those investigatory skills. And add on to those skills what is necessary [...] for him to understand telecommunications and you've then got a telecoms investigator. So, by virtue of that, you will find a great deal of law enforcement coming into the industry as opposed to the other

way around. In other words, the flow tends to be one way (Public Affairs Officer, MI5, Interview 33 2000, pp. 543–552).

There were early signs, however, that under RIPA and CALEA, the relationship between the telecommunications industry and law enforcement would assume an increasingly more structured and defined format. This process was initiated through the negotiations between law enforcement and industry that preceded and followed the implementation of digital communications interception legislation. In the United States, most such negotiations were facilitated by the Electronic Communications Service Provider Committee (ECSPC), an industry-law enforcement forum created in 1992 specifically to address digital barriers to traditional models of state-sponsored communications interception. The forum, which held its first meeting at the FBI's Quantico, VA, facilities sometime between March and May 1992, operated under the auspices of the Alliance for Telecommunications Industry Solutions (ATIS), and became part of Telecommunications Industry Association Standards Forum, following the implementation of CALEA (Morris 1998 n15; Meeks 1994). Notably, these negotiations were unsuccessful and convinced the United States law enforcement and security community that the telecommunications industry was not willing to voluntarily facilitate communications interception. This realization led to the emergence of CALEA. However, the FBI later acknowledged that the ECSPC served as the platform for "literally hundreds of meetings" between law enforcement and industry representatives (Kallstrom 1997; see also Freeh 1997).

In the United Kingdom, three groups were active as channels of contact between law enforcement and industry actors engaged in negotiations over the RIPA. By far the most influential of those was the Joint ACPO, ISP and Government Forum, which preceded RIPA's enactment by more than 2 years. Ever since its formation, the forum held regular meetings with the objective to

> develop and maintain a working relationship between the ISP industry and law enforcement agencies in the United Kingdom, such that criminal investigations are carried out lawfully, quickly and efficiently while protecting the confidentiality of legitimate communications and with minimum impact on the business of the Industry [Pearson 1998; see also HCTISC 1999a].

The ACPO Police and Telecommunications Industry Strategy Group was another forum that served as a platform for negotiation between law enforcement and wireless industry representatives concerning RIPA (Anonymous 2000b). In a leaked report, NCIS Director Roger Gaspar asserted that, through the ACPO and Telecommunications Industry Strategy Group, industry and law enforcement actors "developed a strong strategic partnership" (Gaspar 2000, Sect. 4.1.3). Parallel meetings between the law enforcement and security community, on the one side, and ISPs on the other, were facilitated through the Internet Crime Forum (ICF), a group that aimed to "develop and maintain a working relationship" between the two sides (Perry 2000b; see also Perry 2000a; Interview 10 2000, pp. 391–395).

All legislative and practical accomplishments aside, the above groups served as vital channels of recognition and mutual identification of individuals on both sides, who were instrumental in creating and implementing the operational and legislative communications interception models incorporated in RIPA and CALEA. According to one ICF member, the fundamental meaning behind these negotiations was that

> if people need to get hold of one another, there is quite a lot of recognition—if you like—between people like myself and law enforcement people. So, if you needed to go and get an opinion about something to do with law enforcement [. . .] you can just [. . .] pick up the phone and talk to them [Law Enforcement Liaison, large British ISP association, Interview 29 2001, pp. 397–398, 401–405; see also Watson 2000];

> I guess the good thing that came out of the ECSP[C] meetings was a networking of contacts within law enforcement [and] within the telecom industry, that followed through for the next few years as people would call [and] contact [each other] once CALEA was passed [FBI Special Agent and Telecommunications Industry Liaison, CALEA Implementation Section, Interview 40 2000, pp. 72–75; see also HCTISC 1999a].

The enactment and gradual implementation of RIPA and CALEA was also expected to facilitate common educational seminars and training schemes between law enforcement and TSPs:

> we'd certainly be looking to do more [common training]. And there [are] lots of avenues to achieve that by Telecommunications UK Fraud Forum (TUFF). There's also an annual conference [that the Federation of Communication Services] and TUFF run for law enforcement and European law enforcement, so there [are] lots of opportunities there [. . .] to network, plus to [. . .] do the training side [Fraud Control and Revenue Assurance Manager of large United Kingdom wireless communications company, Interview 04 2000, pp. 198–202];

> [y]es, [. . .] we'll be seeing much more of them and they'll be seeing much more of us. It's been like this for a few years, but now it's going to be more specific [. . .]. It's going to be less [of] 'are we going to implement CALEA?' and more of '[. . .] what's the most feasible way [to] do intercepts under CALEA?' [FBI Special Agent in Charge, Interview 35 2001, pp. 222–227].

Undoubtedly, therefore, the institutional interface between state agencies and TSPs was expected to increase under CALEA and RIPA. This was not a source of excitement for either of the two parties, but both recognized that they were forced to cooperate by technology and legislation.

6.9   Exchanging State Monopoly for Chaos

Admittedly, one of the most fortified digital barriers against communications interception has not been technical, but rather regulatory. In the closing decades of the twentieth century it was recognized that the disappearance of large, state-owned and state-sheltered telecommunications monopolies, which occurred in the 1980s, was directly responsible for an equally dramatic breakup of the once-compact community of law enforcement and telecommunications security officials. In the United States, the process began in 1980 and had been completed by 1991 (Stone 1989, p. 11; Cohen 1992, 3ff). In the United Kingdom, the process was initiated in 1981 and had been completed by the summer of 1993, when the British government reduced its British Telecom holdings to nearly zero, so that British Telecom became in effect fully privatized (Spiller and Vogelsang 1996, p. 95). The astronomical increase in the number of telecommunications carriers and service providers in the United Kingdom—by 2000 more than 200 PTOs and several hundred ISPs, according to estimates (Anonymous 2000b)—virtually brought certain areas of state-sponsored wiretapping to a halt (Interview 02 2000, pp. 121–124). Indeed, United Kingdom ISPs were known to aggressively use the absence of any meaningful regulation of their industry as a weapon against the government during negotiations over RIPA. In 1999, one ISP representative wrote in a government consultation document that "we don't need a license to operate our network, so blackmail in this case will simply get you unwanted publicity" (Parr 1999; see also Interview 05 2000, pp. 533–539; Interview 08 2000, pp. 62–64ff). British carriers were certainly aware of the tension that had been caused by deregulation:

> [w]ithout doubt, it is a huge problem. Quite often [we] may be the second or third call that a policeman has made trying to identify who owns a particular [Calling Line Identifier] or complete service operator. So, sometimes I wouldn't envy their position—they just can't locate who the service provider is now [. . .]. It's [. . .] a problem [and] I am sure it is going to get worse in the future [Fraud Specialist, large British telephone service provider, Interview 24 2001, pp. 208–211, 220; see also Interview 07 2000, p. 227; Interview 10 2000, p. 468; Interview 12 2000, pp. 369–370];

> [law enforcement] would probably start with [. . .] the British Telecom books that say CLI ex—this geographic region—and this is a number owned by[, say,] Cable & Wireless. So they phone Cable & Wireless [and] Cable & Wireless say 'no, this has been passed on or ported on to', say, 'Energis'. They phone [Energis] and then [they] say 'well, actually, we've actually put this on through an [. . .] account and you need to phone this number'. So [. . .] at minimum you're talking about three or four calls. So it can be a bit of a problem' (Interview 24 2001, pp. 213–218).

In the United States, law enforcement officials did not hesitate to describe the deregulation of the country's telecommunications sector as "dramatic" (Yarbrough 1999), "vastly complicating" (Boucher and Edwards 1994) and "chaotic" (Blair et al. 1995) from a security viewpoint. Meanwhile, industry representatives asserted that the sheer number of new TSPs in the deregulated market meant that agents in the United States government were "looking at a nightmare in administrating CALEA" (Donald Bender, United States Telecom Association's CALEA Officer in Auble and Bender 2000). Indeed, law enforcement interviewees in the United States overwhelmingly confirmed the seriousness of these fears:

> [t]he dynamics [. . .] of the thing are so varied—and if they are not directly related to CALEA itself, they're related to the number of players that are now in the telecoms world [. . .]. The paradigm broke [with] the [. . .] Telecoms Bill of 1996, which allowed every entrepreneur to enter the telecoms market. I just cannot convey to you the volume of people, who know nothing about telecoms, that have jumped into the market. And, as a result, everything has been bought up, absorbed or split up and there are so many players. The vast majority of the people now are not sure what their responsibilities are. [And] it's not just a dozen carriers we [. . .] are working with; there are 37,000 carriers. Half of the carriers today did not exist in 1994. It's that big [. . .] change [FBI Special Agent, CALEA Implementation Section, Interview 19 2000, pp. 757–761, pp. 806–809, pp. 817–819; see also Interview 20 2001, pp. 155–158, 205–207; Interview 21 2001, pp. 144–146];

[b]efore the breakup [...], say, five years ago, it was all local people that I dealt with and it was all people that [we] knew [...]. And they [...] could help me at two o'clock in the morning as well as they could at ten o'clock in the morning. Because they knew who I was, they didn't have to get clearance. After [the carrier] relocated—see, they got bought out—[...] to North Carolina, a lot of times when we have problems and call in they don't even know [who and] where you are. They want to know where you are [Technical Unit Chief, Emergency Communications Unit, state government agency, Interview 13 2000, pp. 145–151].

You've got [...] fourth parties now. Here we've got one supplier that is buying his lines from a guy that bought them from Sprint [...] So when we have a problem with a call and we go back to Sprint, Sprint says 'well this is Rogers Communications'. So we thank them and we call Rogers Communications and they say 'oh yeah, we'll help you out. Oh, I'm sorry, we sold that to Larry Communications. We don't have any record of that'. Who is Larry Communications? 'Oh, well...well, I'm sorry, I can't tell you that; that's confidential information. You have to get a court order and I'll give it to you'. And, I mean, this could take forever. You've got to get a court order. You've got to find a judge, you've got to prove to a judge that you need [...] this information and he's not going to give it to you because of who you are and who you work for. He's not going to do anything, because he doesn't want to get in the middle of a lawsuit. So I've got to prove to him that I've got to have it. And [...] this has got to happen quick, because there might be a man that has called in because somebody is in the other room killing his family and we can't figure out where he is at. We can hear all this commotion going on and we start tracing and we just run into dead ends. We didn't use to have that. There were no third parties [...]. Once it almost cost us dearly; and I say 'almost' because at the last second somebody [from the carrier] went around all the red tape and said 'listen, I believe that if you go out to Joe Smith's you'll find that that's where this phone goes'. A few times we've done that. It was a real emergency and nobody went back through the channels to figure out how we knew what we knew. They figured it's just police emergency, everybody thinks they know that we know where everybody's at. So nobody checked. But we don't know where they're at (Interview 13 2000, pp. 108–123).

It is important to stress here that this new reality in the deregulated telecommunications landscape was completely uncharted for both state agencies and service providers. The legislation that came in its heels to facilitate communications interception, namely CALEA and RIPA, did not necessarily introduce order into a chaotic situation. It merely forced state agencies and carriers to talk to each other so as to devise new procedures through which to facilitate communications interception in the new, digial telecommunications environment.

6.10 Communications Interception in a Global Deregulatory Environment

Law enforcement and industry officials also alerted legislators to the fact that global deregulation developments in the field of telecommunications could potentially mean that communications interception facilities could be based outside the country where state-authorized communications interception was sought. Hence, nationally valid-warrants could be increasingly rendered invalid in a rapidly global telecommunications environment:

[i]t is vital [. . .] that a [carrier] company [. . .] maintain within the United States interception access to the stored wire and electronic communications of their United States customers and subscribers and any records and subscriber information relating to such United States customers or subscribers. If such information is unavailable because it is stored beyond the US' borders, subject to restrictive disclosure laws of foreign countries, or technologically inaccessible, the national security, law enforcement, and public safety interests of the United States are degraded proportionally [STTCPUSHRCC 2000; see also Blair et al. 1995];

[y]ou know what [could happen]? [Carriers] can put the [switches] in another nation. And that's a real. . .now you're jumping not only to technical problems, but [also] to legal problems. Think of this: 'why don't we put [them] in Tijuana, Mexico? Labour's cheaper, it's pretty close to our copper'. Now, could I have gone down to Tijuana with a US-issued court order [. . .] for a drug case? Oh, man, they're going to laugh me right across the border. So, that's creating a lot of heartache for law enforcement [FBI Special Agent, CALEA Implementation Section, Interview 19 2000, pp. 641–647];

Demon Internet already operates a network with equipment in three countries and has plans to operate in further parts of the European Union. Some parts of the user access infrastructure for the Netherlands are in fact physically placed in the United Kingdom and at times United Kingdom infrastructure has been housed in the Netherlands [...]. Is a United Kingdom TSP expected to act on a warrant when the best place to place the intercept is physically located in the United States where wire-tapping legislation is rather different? If the government sees this as a design issue ('you should not design your networks that way') then this is a considerable imposition on our freedom to develop our business in an effective manner [Hall et al. 1999; see also Parr 1999; Emery 1999; Sutter 2001].

This is probably the area of CALEA and RIPA that highlights the most serious long-term concerns for national security. The globalization of the telecommunications market has only increased since the introduction of those two pieces of legislation. At the same time, the fragmentation of the business model in telecommunications has not made the process of cross-border cooperation easier. The law has found it difficult to address deregulation within a single and unified nation's territory. One can only speculate as to the difficulties of doing the same across national borders or even across continents.

6.11 The Cultural Shift of Deregulation

In addition to the legislative shift, which resulted from deregulation, United Kingdom interviewees indicated an equally significant—yet largely implicit—cultural shift that characterized some of the underlying qualitative differences between former state monopolies and the new, private-sector TSPs. Both British Telecom and independent service provider interviewees appeared to recognize the special status that British Telecom enjoys in the industry's relationship with law enforcement. One regulation officer of an established British TSP remarked:

there's a gap [. . .] between ourselves [and new carriers]. British Telecom [has been] carrying out interception on voice networks [...] for years; there are established processes and procedures and established cost-recovery mechanisms, so that we can recover part of our reasonable costs (Regulation Officer, large British TSP, Interview 31 2001, pp. 118–121; see also Interview 07 2000, pp. 251–255).

I think it's [. . .] different in a lot of ways, because British Telecom have their own investigations unit and investigation arm, which [. . .] means that they can talk to [law enforcement] in the same sort of level and they have a lot more powers [. . .] from their [. . .] sort of [ex-]Post Office days. And they do tend to sort of take a lot more ex-police and ex-[secret] services anyway. So they talk the same language. And [it's] also their size—[. . .] they have a lot more power. With ourselves, [. . .] we have to do it by friendly social beings [Fraud Control and Revenue Assurance Manager, large British wireless communications service provider, Interview 04 2000, pp. 224–231];

[t]he Post Office has been assisting in the interception of mail and then telephone calls [. . .] since the 17th century. So, if you like, the [. . .] culture of co-operation with the government and the [. . .] maintenance of whole structures within [. . .] British Telecom that are concerned with the interception and maintaining the secrecy and security over interception operations, has been there organically, sort of stripped through the organization. And that cuts both ways; I mean, not only does it mean that British Telecom are very efficient in doing that and they have specially-trained [. . .] engineers who do that [. . .], but that the govern- ment trusts them to do that job [Bowden 2000, pp. 555–563];

[there is considerable] difference between different companies and their attitude to what extent their organization is part of the law enforcement process or [. . .] to what extent their organization is incorporated to law enforcement. I think the older the institution is and the more established it is, the more it feels that [. . .] it's part of the [law enforcement] establishment [Law Enforcement Liaison, large British ISP association, Interview 29 2001, pp. 222–227].

In contrast, the institutional interface between law enforcement and newer TSPs remained largely unstructured and unformed, something that was typically reflected in their cultural structures:

ISPs [ha]ve only sprung up in the past seven [or] eight years. Before that there wasn't such a thing as an independent service provider. [C]ulturally they have few [. . .] commonalties with law enforcement [. . .]. Whether the[ir] relationship [with law enforcement] can ever be quite as sort of cozy and seamless as it has been with British Telecom, I doubt [Bowden 2000, pp. 563–566; 602–603];

there is [. . .] an agreement [between British Telecom and law enforcement] that there's [. . .] a need for interception, which is legitimate—a need to meet certain objectives in society. Where there's some more doubt and less understanding is [. . .] typically with ISPs—the small operators—who have not been involved in the process at all before [Regulatory Affairs Officer, large British TSP, Interview 11 2000, pp. 80–83].

Law enforcement officials appeared also to be recognizing this cultural shift. Prior to RIPA's enactment, they consistently called for a change in the legal regime of communications interception in order to limit or even eliminate the potential harm to state-sponsored communications interception by uncooperative TSPs:

[w]hilst we currently enjoy a level of co-operation from most communications service providers unparalleled in Europe and the rest of the world, this cannot be guaranteed and, as such, th[e government's] proposal to introduce a statutory requirement appears prudent [Sims 1999; see also Veigas 1999; Maddison 1999].

The situation appeared to be rather similar across the Atlantic, with older carriers revealing a sense of reassurance about their historical arrangements of facilitating state-sponsored communications interception. But younger carriers were often refusing to honor communications interception requests in the absence of sufficient documentation. Consequently, law enforcement reminisced about the "good old days" (Interview 19 2000, 723ff) of the AT&T monopoly:

our company has been doing intercepts longer than I have been alive [and] are fully aware of the security concerns. [As an] incumbent carrier [our company] has had a long-standing relationship—a long-standing history of co-operation—[with law enforcement] and the younger carriers haven't even thought of what they have to do. They're [. . .] not as [. . .] mature a company as we are. I am sure that when they are presented with a court order [. . .] it's a new thing to them and they're not used to dealing with that sort of thing [CALEA Compliance Attorney, large United States TSP, Interview 20 2001, pp. 321–327];

[our relationship with law enforcement] is just like anything, you know, you build rapport and then you have a certain comfort level at some stage in the relationship [CALEA Administrator, large United States TSP, Interview 41 2001, pp. 155–156];

[we have] policies and procedures on how to handle a request [. . .] for a surveillance and [. . .] we're going to stop them at the door and say 'hey, this is what you need to have and this is what we're going to do. And if you don't have it you ain't gonna get it!'. Bottom line [Policy Lobbyist, national association of rural TSPs, Interview 38 2000, pp. 229–233];

before the break-up, [. . .] the phone company [. . .] was totally responsible. They'd give you that information within as quick and orderly fashion as possible [Head, Emergency Communications Unit, state government agency, Interview 34 2000, pp. 123–126];

the industry, 20 years ago, was half a dozen companies. And I knew all of the players and the security people. And they had a very tight ship as far as the security of the information; and it was a [. . .] very closed family, this relationship between law enforcement and telecoms. Not so any more. Telecom carriers in particular are springing up every day. New people, companies that have never even thought about the intercept angle that they're going to be involved in. So, in that respect, there are a lot of new players many of whom we do not have close working relationships with and they may not share [our] main concerns about keeping the information secure as we would or other carriers [would]. I know that sounds goofy but I would view [. . .] and characterize [the pre-break up era] as the 'good old days'. It was easier, it was old-hat, [. . .] everybody had done it. It was pretty easy and straight-forward [FBI Special Agent, CALEA Implementation Section, Interview 19 2000, pp. 681–689, 723–725].

Arguably, the technical knowledge gap between established and new TSPs would decrease with time. However, this did not help to ease the sense of uncertainty felt in the ranks of law enforcement professionals, who found themselves having to learn to interact with new companies who had a very different corporate background and culture.

6.12 The Financial Shift of Deregulation

Apart from encouraging a variety of cultural shifts within the telecommunications industry, deregulation also entailed financial implications for carriers. The degree of competition escalated, transforming the United Kingdom and United States telecommunications markets into symbols of antagonistic liberal market ideology in late twentieth century global capitalism. Consequently, law enforcement's communications interception needs were subsequently degraded into low-priority concerns by an industry that was simply too busy being competitive:

> the newer, brasher [telecommunications carriers a]re rushed off their feet just doing their day job and [to] anybody who comes along asking for something else they're going to say 'oh, heavens above, I'm far too busy, go away! I'm trying to run a business here' [Law Enforcement Liaison, large British ISP association, Interview 29 2001, pp. 480–484];

> [carriers had] heard what law enforcement's needs [and] requirements were—they'd had that in writing—and, at a technical level, they understood it. But they were not willing to make any modifications to their equipment, because [. . .] they had their most talented engineers working on features and modifications to the equipment that would make them money. Why work on this non-profit making feature? As important as it was to us it was certainly [not] bringing in anything for them [FBI Special Agent and Telecommunications Industry Liaison, CALEA Implementation Section, Interview 40 2000, pp. 168–174];

> law enforcement are looking to us to volunteer to put [CALEA technology in everywhere at our own cost [. . .]. But in a competitive environment we can't give stuff away, you know, we've got to operate to make a profit [CALEA Project Manager, large United States TSP, Interview 21 2001, pp. 96–98].

Industry representatives were particularly vocal about what they perceive to be a direct and threatening link between, on the one hand, optimizing RIPA and CALEA modifications—described by a senior British telecommunications executive in 2001 as an "intolerable commercial burden" (Delio 2001)—and, on the other, maintaining the financial prosperity of their business:

> [in implementing IOCA] the government runs a very serious risk of damaging our ability to compete [. . .]. Any proposal that degrades the level of service that we can offer to our customers will affect our position in a highly competitive worldwide market [Hall et al. 1999; see also Perry 1999; Anonymous 1999d];

> [f]orcing industry to become CALEA compliant in under two years would not serve 'the policy of the United States to encourage the provision of new technologies and services to the public', as enormous amounts of time and engineering manpower—otherwise employed in the provision of such desirable technologies to the public—would have to be dedicated to an accelerated implementation of CALEA [Baker et al. 1998].

Rather predictably, therefore, concerns relating to financial compensation by government for CALEA and RIPA network modifications topped the industry's list of concerns over the two pieces of legislation, far surpassing information security or capacity issues (Interview 01 2000, pp. 197–199, 210–214, 359–361; Interview 19 2000, pp. 553–562; Evans 1999):

[financial compensation] is an overhead. What's in [assistance to law enforcement] for any organization? You're a good corporate citizen, you're a good member of society, if you like, but it's not actually adding to the bottom line [. . .]. We are looking at [RIPA] for commercial reasons, actually, rather than particularly moral reasons [Regulatory Affairs Officer, large British TSP, Interview 11 2000, pp. 63, 106–108];

in the wireline world [carriers] go to these commissions and make a plea to modify their rate structure to recoup these [CALEA modification] costs from their subscribers. No carrier wants to do that. Who wants to raise their rates? Who wants to put on a 'wiretap tax'? It would kill them. So that's not really a viable option for many carriers. Well, wireless people, they're not in any better position. If you look in today's paper you'll see them begging you, giving you minutes or extra time for a dollar less, everybody's cutting each other's prices. Well, the last thing they want to do is buy more expensive equipment to provide you with that same service. They want to do just the opposite. So, [CALEA] capability equates to cash. The less capabilities they provide, the more cash they're able to provide their stockholders (FBI Special Agent and Telecommunications Industry Liaison, CALEA Implementation Section, Interview 40 2000, pp. 553–562).

The first thing the phone company says when you meet with them is 'will we have to charge the customer under the [CALEA] contract?' (Head, Emergency Communications Unit, state government agency, Interview 34 2000, pp. 11–12)

I think that the breadth and the scope of our putting the CALEA functionality in our network is going to depend upon the extent to which we can expect reimbursement of our costs. And so, [...] the extent [to which] the government can [. . .] identify additional moneys in other areas of law enforcement—that will help facilitate CALEA [CALEA Project Manager, large United States TSP, Interview 21 2001, pp. 173–175].

Increasing sensitivity over public relations—arguably a side effect of customer-driven deregulated industries—was also influential over the industry's overall view of CALEA and RIPA, although apparently it often had contrasting effects:

you see, it's not [. . .] that there may be [a] more civil libertarian sort of attitude in the management [of certain carriers] or whatever. It's just the fact [that], if they get things like that wrong, they might end up being sued by a customer for damages [. . .]. And if it leaked out that a particular ISP was over-co-operative with the police on interceptions, a lot of the customers would say 'I'm sorry, we're going to find another ISP' [Regulation Officer, large British ISP association, Interview 10 2000, pp. 583–589];

[carriers] were very much [thinking]: '[what would happen] if we sort of had a [public relations] scandal, if something like this happened and our network, our company, our equipment was identified as the problem by which the little girl was kidnapped because law enforcement couldn't get her contact information'? They were very scared of that. [A]nd that was the only motivator that I saw. I did not see any altruistic motivation by the carrier or the manufacturer to take [. . .] a stand and do this because it was the right thing to do. I'd love to say that [FBI Special Agent, CALEA Implementation Section, Interview 19 2000, pp. 183–192; see also Interview 13 2000, pp. 140–142].

Admittedly the end of the telecommunications monopoly in both the United States and the United Kingdom came with a renewed requirement to compete for the 'hearts and minds' of consumers. Service providers therefore became very quickly extremely concerned about their public image and the need to be seen to protect their customers' privacy from unwarranted intrusion.

6.13   Negotiating Over Digital Communications Interception

Any attempt to outline the negotiations that preceded and followed the creation and introduction of RIPA and CALEA should commence by attesting to the element of secrecy, which has inevitably withheld large portions of the debate from researchers and non-security-cleared observers.

Evidently, to representatives of United States governmental organizations, large segments of the debate over CALEA were seen as a matter of the citizens trusting their elected representatives to handle security issues that are too sensitive to be publicly negotiated. These convictions helped erect near-impenetrable security barriers around some of the more crucial aspects of the digital communications-interception debate. Asked at a press conference whether American citizens should trust the nation's intelligence services in having a say in the design of the digital telecommunications infrastructure, Michael Nelson, who at the time was the United States government's advisor on the National Information Infrastructure, laconically replied: "[i]f only I could tell you what I know, you'd feel the same way I do" (Barlow 1994). Another supporter of law enforcement's demands has remarked that

> [u]nfortunately, it is not possible for most of us to be fully informed of the national security implications of [these matters]. For very legitimate reasons, these cannot be fully discussed and debated in a public forum. It is even difficult to talk about the full implications [...] on law enforcement. This is why it is important that the President and Vice President be fully informed on all the issues and for the decisions to be made at that level [Denning 1994].

These opinions largely reflected attitudes on communications interception among the nation's security and intelligence establishment. In a lecture delivered at the Massachusetts Institute of Technology, former Director of the NSA, Deputy Director of the CIA and Director of the Office of Naval Intelligence Admiral Bobby Inman admitted that, had the NSA had its way, the current policy debate on digital communications interception would not have "over-spilled" onto the public arena (quoted in Lethin 1995). A few months earlier, Federal Agent Barry Smith, a member of the FBI's Congressional Affairs Office, had remarked that, if a solution to the digital communications interception problem was eventually reached, the FBI would avoid even publishing a press release on it: "it will be done quietly, with no media fanfare', he admitted (Meeks 1994).

Regrettably from the standpoint of scholarly research, large—and arguably the most crucial—segments of the negotiation process that eventually led to the introduction of RIPA have been clandestine. The initial interactions between law enforcement and industry representatives, in which a solution based on consensus, rather than on legislation, was pursued, were conducted in absolute secrecy. According to Professor Yaman Akdeniz, who first exposed the existence of the initial set of meetings,

> we thought it was secret—which it was. They called it an 'internal document', [but] I think it was secret. It's pretty safe to say that was secret, because once all interested parties are not

included in these meetings or negotiations, then there is secrecy [Akdeniz 2000, pp. 376–378, 24–27; see also HCTISC 1999a].

Following the initial introduction of IOCA, when the negotiation process over RIPA had a less clandestine format, a high degree of secrecy remained a foremost precondition for the debate, and it was apparently respected by all participants, including civil liberties groups:

[w]hat generally happens after several meetings [...] is [that] some sort of trust develops [and] a dialogue of some kind is going to be possible without necessarily everything that either side says being [. . .] noted down—i.e. put on the Internet or trumpeted to newspapers [Bowden 2000, pp. 296–300];

we [all] respect [each] other's confidence and certain information handed over. I mean, some of the information is clearly public. So, for example, when this civil servant gives me—as he did—[. . .] a copy of this and says 'here you are, have a copy of the Smith Report and read it' [. . .], he will give me a copy of this two days before it's published. We [. . .] went to the Home Office and they said 'look, here is a copy of it, please don't show it to anybody, but there you are'. So [. . .] the relationship of trust and good understanding may well be coming from why we've been invited to the meeting—what we're supposed to do, how we're supposed to react [Regulation Officer, large British ISP association, Interview 10 2000, pp. 169–177].

To a large extent, the secrecy that permeated negotiations over RIPA was justified by the sensitive nature of the subject under negotiation:

the whole process has been up to now fairly restricted to [. . .] telecoms operators and [in] the existing IOCA it's to be restricted [. . .] to those. So, [it's] been quite a homogenous population of people—and a small population—who have been required to intercept communications on behalf of law enforcement authorities. So, those people have had a clear view, to start with, of why it's been required and understood the controls that were applied, which certainly previously have not been particularly visible outside of this, sort of, fairly exclusive club [. . .]. The number of people who know the detailed process—the less they are, the better, really. [I]t is a difficult issue [. . .] simply because previously [. . .] it's been [. . .] a very specialist area, which, [. . .] by its very nature, you tend to need to keep secure. You don't want to [. . .] talk about it too much, because it's counterproductive [Regulation Officer, large British TSP, Interview 31 2001, pp. 20–26, 89–92, 125–126].

Nevertheless, British law enforcement and security community interviewees claimed that even the acknowledgement by law enforcement and security agencies of lines of communication with carriers marked an astounding shift in policy, namely the open recognition of a relationship that would have hardly been publicly admitted, let alone documented in previous years:

I mean, you've only got to go to the NCIS web site to see how open that organization is. And certainly, I mean, you know, in the timeframe of the last two to three years, increasingly more open. [P]rior to that, no. Prior to that, [contact would take place] behind close doors. And you wouldn't even discuss it—you wouldn't even [. . .] admit its existence half the time [Officer, MI5, Interview 12 2000, pp. 451–455];

I can tell you for a fact [that] some [of the] official documents [on the RIPA that are] circulating in public shock even law enforcement [members] themselves—some of the older members. I'm sure they're shocked to see no 'restricted' [labels] on these documents. I am sure they are, because I am too. But times [are] changing [. . .] rapidly [Communications Intelligence Officer, MI5, Interview 02 2000, pp. 319–321, 344–345].

In reality, despite the alleged "over-spilling" of the communications interception debate into the public domain, very few details are known even today about the negotiations between law enforcement and industry, while even less information is available about the discussions between law enforcement and government officials prior to the introduction of CALEA. Throughout the 1990s, numerous closed-door briefing sessions were conducted between members of the United States Congress and members of American law enforcement agencies, including FBI Director Louis J. Freeh. In fact, one such session, held in June 1997 by the House International Relations Committee, raised vocal criticisms by non-governmental observers (Anonymous 1998b; see also Leahy 1995; Berman and Dempsey 1996). In general, the vast majority of governmental, law enforcement and industry representatives who took part in CALEA negotiations required security clearances and usually received a *Regulations and Procedures* memorandum warning them not to reveal classified information imparted to them, not to make written notes of classified discussions, and to report to the authorities any attempt to obtain classified information from them (McCullagh 1998).

6.14   Law Enforcement Disunited

Much of the contemporary and historical analysis of the RIPA and CALEA debate has tended to assume that the negotiations over the two pieces of legislation were primarily conducted between two groups of united and cohesive actors, namely the law enforcement and security establishment, on the one side, and the TSP community, on the other (Sutter 2001). A more thorough examination, however, shows that these two groups were characterized by the unequal and problematic alignment of disparate parties in pursuit of common interests at its simplest form, or by the patchy amalgamation of competing interests at its most complex. The British law enforcement bloc was a particularly interesting example of that. Although official documents appeared to show uniformity of opinion over RIPA among ACPO, HM Customs and Excise and the various intelligence agencies, including NCIS, MI5 and GCHQ (Gaspar 2000, Sect. 1.1.1), there are reasons to believe that this uniformity was, in fact, not reflective of reality. According to one interviewee:

[e]ven though the ACPO people are the most noticeable law enforcement people out there in defence of [the] RIP[A, the] National Criminal Intelligence [Service (NCIS) is] the real engine behind the campaign [. . .]. You have got to remember that ACPO is made up of older policemen [sic]—old-school police [officers]—who don't really understand much about electronic interception. I mean, very few Police Forces out there make heavy use of interception in investigations: Scotland Yard, Greater Manchester, West Midlands, maybe Glasgow and Leeds [. . .] and that's [. . .] about it, really. So [ACPO] would have had limited knowledge about the [communications interception] problems that came up [. . .]. Maybe some people in CID did encounter some [problems]—in fact, I am sure they did—but they

would have had no way of knowing why the[se problems] were appearing. So the leads came from NCIS and [. . .] GCHQ and were eventually streamed all the way down to ACPO [. . .]. And, frankly, [. . .] most of the ACPO people are not particularly bothered about [the] RIP [A], because very few of them will ever use it. NCIS, on the other hand, are very up-front about some of the more technical issues [in the Act], such as encryption or data retention, etc., because they will be the ones that will primarily use them—they and Government Communications [Headquarters] [Officer, National Criminal Intelligence Service, Interview 22 2001, pp. 411–413, 417–430].

Some of these differences of opinion, which unavoidably existed within the law enforcement RIPA lobby, and perhaps within ACPO itself, can be traced to the written responses to the Home Office by law enforcement officers responding to the IOCA consultation. One respondent called for a more open discussion on communications interception, disputing the official law enforcement argument that detailed public discussion over the issue would benefit criminals:

I do not agree entirely with the argument that exposure of interception capabilities will educate criminals as those involved in crime at this level are already well aware of what our capabilities are [Stoddart 1999].

The same respondent rejected a proposal to have blanket communications-interception authorization protections from the NCIS, and requested instead that local police forces authorize their own communications interception requests without the need to resort to NCIS interception experts (Stoddart 1999). Another respondent distanced himself from the NCIS' view of the European Convention on Human Rights (ECHR) as a threat to British law enforcement's interception capabilities and stressed that

[c]ompliance with ECHR is of paramount importance and this must be stressed throughout the consultation process [Grey 1999; see also White 1999].

Two other respondents expressed the view that communications interception requests to TSPs should be kept at an absolute minimum and be reserved for cases of serious criminal activity with ample consideration given to the intrusion that the practice of communications interception can represent to the private lives of citizens:

[c]ommunications service providers cannot be expected to supply information to the police for every case, no matter how minor [Todd 1999];

[t]here are obvious concerns as to the consequent level of intrusion, which [communications interception] represents and I would suggest that this dictates the minimum authority level to be at Detective Superintendent Rank (as currently exists in my Force). This should provide a safeguard to those concerned with the potential for abuse and reassurance to the data providers that their interests are being considered [. . .]. The principle must be that we are 'whiter than white' and must be able to evidence that fact [White 1999].

In the United States, the law enforcement community's federalized character meant that the law enforcement lobby was even more fragmented than the United Kingdom's. Almost a dozen federal bodies were actively involved in CALEA lobbying and implementation, including the United States Secret Service, the Marshals Service, the Postal Service, Alcohol Tobacco and Firearms (ATF), the Drug Enforcement Administration, the Internal Revenue Service and, of course, the FBI (see Yarbrough 2000). Many of those organizations were often in disagreement about the specifics, much less the general direction, of the legislation and its implementation:

> when the [J-STD-025 or] J-Standard came out, in December of [19]97, [things] turned really sour. I mean [...], it was almost like we had never even talked to the industry—they [...] had just gone out and [...] done their own thing. We had [...] to go up to Chantilly [and] it was just...simply unpleasant. Allegations were flying across the room; some [people] questioned the FBI's handling of this whole CALEA operation. There were two [representatives] there from the Secret Service and they [...] eventually [...] got up and left the meeting. It was that bad [FBI Special Agent in Charge, Interview 35 2001, pp. 112–113, 115–122];

> [n]ow, [...] I know that [the FBI] have formed a law enforcement group [or] consortium [...]. When dealing with them on some of the issues [...] they seem to be saying 'OK, we agree with you' and then they would disappear and come back and say 'oh, we can't agree with you'. Because they would get together with th[e] DIA and...and the CIA [...] up there in New York City [...] with a whole bunch of law enforcement people and [...] go through this [Director, National Services Planning, large TSP association, Interview 37 2000, pp. 260–266].

Virtually none of the American interviewees who were actively engaged in CALEA negotiations were prepared to interpret law enforcement's CALEA lobby as a solid and uniformed bloc of actors. That was especially the case with industry representatives, who appeared to be well aware of cross-departmental conflicts between government agencies:

> [t]o do a surveillance, [local law enforcement agencies often] have to contact the FBI in order for them to get the information [on their behalf]. And they don't like that. People don't like working with the FBI, because they find that the FBI comes into their jurisdiction and they kind of take over. And then you start getting into turf battles (CALEA Lobbyist, large TSP association, Interview 16 2000, pp. 272–276).

> [t]he distinction between the way law enforcement is tiered here in the States [means that] we have a lot of turf battles. And everybody guards their territory rigorously and they don't like anybody exceeding over into their [...] territory. And we constantly [see] that in the federal agencies. You've got [...] the Customs, Director of Central Intelligence, FCC, [...] Secret Service, ATF [...], and all of them go up against the [Federal] Bureau [of Investigation]. Because the Bureau has [...], I would say, almost omnipotent authority [...]. Their jurisdiction overlaps into a lot of all the other [agencies'] jurisdictions, even though the others have primary [authority] in certain areas [...]. But that's always been like this in law enforcement. There's always been conflict. When you cross turf [it's like] 'that's my boundary, don't cross over. If you do [it] you'll get my wrath' (CALEA Project Manager, large United States TSP, Interview 21 2001, pp. 76–87).

Although it was generally agreed that the FBI had been the government lobby's primary representative during the negotiations (Interview 20 2001, pp. 29–30; Interview 16 2000, pp. 267–270), it was noted that the FBI's predominance was not decided overnight:

> you've got a bunch of different federal agencies that have surveillance authority and [...] it took a while, I think, for the FBI to be viewed as the lead agency on it. It just took a while for consensus to develop [CALEA Compliance Attorney, large United States TSP, Interview 20 2001, pp. 199–201].

A particularly deep schism that run through law enforcement's CALEA lobby was, on the one side, between Federal and large metropolitan state and local law enforcement institutions, and, on the other, small-sized or rural police departments. According to FBI's own sources, less than a third of American states were actively consulted and represented in law enforcement's CALEA lobbying and implementation bloc (Yarbrough 2000). An FBI interviewee asserted that state and local law enforcement officials were often not even aware of CALEA and its implications:

> [are] state and locals [. . .] aware of CALEA? Well, some are not. Many are. Some are not aware of CALEA's provisions and the fact that [. . .] times are changing as far as the way intercepts are done. I will not gloss this over. There is a surprising amount of ignorance sometimes [that] I see with state and locals about CALEA and how it's going to dramatically affect [. . .] the equipment that they are going to need, the expertise that they're going to have to have on board and the paths by which [. . .] they are going to do intercepts in the future. They do not have a firm grasp [FBI Special Agent, CALEA Implementation Section, Interview 19 2000, pp. 327–333].

A long-time industry participant in CALEA negotiations with law enforcement confirmed the situation:

> [w]hat's very important to understand is that this is a Federal Bureau of Investigation effort to push this through. Most of our companies who go out to talk to local law enforcement are coming back to us [. . .] saying they have no clue what this is all about. The local [. . .] police departments, the local sheriff's department, the county sheriff's department, even to the point of the state police have no idea what this is about. When you look at the law [. . .] it says 'law enforcement', and that includes all of them, all levels of law enforcement, and yet they have not done the job and brought it down to all the levels of law enforcement as to what this is all about [Policy Lobbyist, national association of rural TSPs, Interview 38 2000, pp. 219–226].

The official reason provided for this dichotomy was that the vast majority of state-sponsored communications interception requests were issued in large metropolitan areas and, therefore, this particular investigative tool does not necessarily concern small- or medium-sized police forces across the country:

> now, where are [communications-interception requests issued] the most? [I]n New York City, [NY,] in Washington, DC, and the major metropolitan areas. And that's not always the voice of [. . .] state and local law enforcement. Knoxville, Tennessee, probably doesn't have, or ever has had, [. . .] an instance where they needed to intercept a satellite telephone. But New York has and the fed[eral agencie]s have too, particularly as we are involved in international criminal conspiracies or across state lines, too (FBI Special Agent and Tele-communications Industry Liaison, CALEA Implementation Section, Interview 40 2000, pp. 316–323).

The actual reasons, however, appeared to be quite different, and had to do with the close nature of the relationship between smaller law enforcement agencies and smaller TSPs. Namely, the personal element that characterized such relationships in smaller communities meant that a considerable number of communications interception requests by law enforcement were authorized on face value by over-familiar carrier executives and employees, rather than by the courts. Consequently, smaller law enforcement agencies had a negative view of the more structured communications interception request format introduced by CALEA, which they perceived to be a threat to their usual—and effective—process of conducting communications interception:

> people are afraid [...] that when you force smaller carriers to build [...] surveillance capacity into their switches, then the small carriers are going to have to charge [...] for the service, [in order] to get their money back—which indeed they will. And so, what you have here is that [local police] forces [operating] on small budgets fear that they're not going to be able to afford to wiretap any more. They're going to have to get charged for it and the whole request will have to be routed through formal channels [by the carrier] [Police Officer with experience in undercover and wiretapping operations, Interview 14 2001, pp. 146–151; see also Interview 16 2000, pp. 270–272, Brown 2001].

6.15   The Industry Mosaic

If the law enforcement bloc in the United Kingdom and the United States was disunited over CALEA or RIPA, then it can safely be asserted that any reference to an industry bloc in these negotiations would be highly misleading. This applied particularly to the United Kingdom, where the formation of a reasonably cohesive industry lobby never occurred throughout the RIPA negotiations. In regards to CALEA, the United States industry landscape was also heavily fragmented, primarily in accordance with the market status of individual corporate actors. Broadly speaking, experienced RIPA and CALEA industry negotiators were keen to stress that "the telecommunications industry is not a single object" (Interview 10 2000, p. 52), but rather a "complex body" (Interview 12 2000, p. 128), various parts of which were "focused on totally different disciplines" (Interview 12 2000, p. 129) and were essentially slow in achieving even basic consensus over digital communications-interception provisions (Interview 20 2001, pp. 183, 183–195). One major reason for that fragmentation was the culture of extreme competitiveness and antagonism that characterized the industry. One British carrier representative characteristically described the way in which highly competitive relations between TSPs affected the industry's IOCA negotiations with the British government:

> when you [...] have these meetings [with the government] it's not always on a one-to-one basis. I think in an informal [one-to-one] meeting [with the government] we'd always [give detailed information]. In the [...] formal briefings that I've been [to] in the Home Office, when you're there with Vodafone, when you're there with Orange [...] or British Telecom, you don't tend to sort of say 'well, we are developing this service and you might not be able to intercept it, because of this, this and this'. Of course the other guys are going to realize what we have in mind and are going to say 'oh yes, that service, how far are you away from launching that? Alright!'. So commercial implementation ties you and you just can't reveal too much, but informally we would give [the government all] sorts of ideas [Fraud Control and Revenue Assurance Manager, large British wireless communications service provider, Interview 04 2000, pp. 282–290].

Another interviewee, from MI5, disclosed that separate segments of the British telecommunications industry operated in almost complete isolation from one another during the IOCA and RIPA consultation period: "telecoms industries didn't even talk to telecoms industries!', he exclaimed (Interview 33 2000, pp. 357–358).

In the United States, similar illustrations of the industry's fragmentation were provided by an FBI interviewee, who held the industry's lack of policy cohesiveness directly responsible for the delay in the standardization and eventual implementation of CALEA's guidelines:

> [u]nfortunately, the ECSPC meetings weren't very productive, mainly because there were competing interests in the room—carriers, manufacturers—all of which were competing against each other, none of which wanted to provide much information as to (a) what information is floating around in the network that [...] may be accessible to law enforcement, [and (b)] how that information could be retrieved. [This was] because they did not want that information revealed to their competitors. And that was a real wakeup call for law enforcement. We had no idea of [...] the cutthroat competitive nature of the industry. And, what appeared to be relatively innocuous information about a telecom carriers' switch, was guarded zealously by the manufacturer—and the carrier, for that matter—and they were [...] very reluctant to provide us [with the] number of switches, the features on those switches, the generic modules on those switches—basically where we would do our intercept [...]. Nothing productive happens when you bring in competitors to a [...] room. [We] got nowhere [by] bringing them together, so we [had to] meet every day with them individually. We had calls with them individually every day about CALEA [...]. And that [...] in itself slowed things down, because you had an assortment of interests in a room—say 60 people, a 'who is who' of the telecoms industry—trying to agree on how an intercept would be done. They had a very difficult time agreeing on how anything was done [FBI Special Agent and Telecommunications Industry Liaison, CALEA Implementation Section, Interview 40 2000, pp. 60–72, 97–100, 795–799].

In the United States, a clearly detectable dividing line within the industry lobby was between switch or network owners—namely the carriers—and switch hardware manufacturers. Specifically, numerous switch manufacturers sided with law enforcement's digital communications interception plans after sensing CALEA's potential for large-scale government-appointed contracts:

> [t]here were some [switch manufacturers] that, for market reasons, decided to be leaders in developing or implementing the CALEA standard and in one case doing it before the standard was complete. [They did this] because they saw a market niche [...] and they foresaw—and correctly so—[...] a demand for CALEA-compliant switching equipment. So [...] they were definitely out of the chute first, making mod[ification]s to their equipment so that they could be there by the market with CALEA-compliant switches [FBI Special Agent, CALEA Implementation Section, Interview 19 2000, pp. 192–198];

> [m]ost manufacturers long ago began designing and developing solutions even to some interception capabilities excluded from the standard. In fact, several manufacturers are well along the way. Moreover, a number of the manufacturers have developed many of the needed CALEA solutions in their switching platforms in order to meet CALEA-like solutions required of them by statute or otherwise by law enforcement or national security entities in a number of foreign countries [Morris 1998].

That was highly reminiscent of the Operator Action scheme, also known as Private Doorbell technology, which was proposed by the telecommunications industry in 1998 and was a variation of key escrow—a short-lived proposal by the American government that hardware manufacturers built telecommunications devices with backdoor access for government agencies. Instead of a third party holding a backdoor decryption key—as was the case with key escrow—Operator

Action allowed the interceptor to access and decrypt encrypted messages at the network's router switch. The scheme, which was approved by the FBI and the NSA, was proposed by 12 hardware and software network manufacturers, which included Cisco, Inc., and Bay Networks (see Mosquera 1998; Mills 1998; Appleby 1995; McCullagh 1998). By late 2001, Cisco and Nortel were already in the process of producing intercept-friendly network switches for both the American and British markets (Interview 11 2000, pp. 132–134), while Lucent, AG Communications, Siemens and Bell Energis had either completed or were in the process of finalizing their CALEA-compliant switching products (Hildebrand 2000; Heupel 1999; Anonymous 1998c). That was done despite strong objections from the service-provider side of the industry.

The industry in both the American and British digital wiretapping debate was also polarized between the older, wealthier carriers, on the one side, and the newer corporate upstarts, on the other. On both sides of the Atlantic, concerned industry participants in the digital wiretapping debate had expressed concern over the logistical costs that would be associated with the implementation of RIPA or CALEA. There were concerns that the requirement of employing staff as points of contact with law enforcement, who would have to be made available to the latter 24 h a day, 365 days a year, could prove particularly costly (Interview 18 2000, pp. 169–172; Perry 1999; Morris 1998; Auble and Bender 2000). At the same time, the labor costs associated with retracing, retrieving and filtering data requested by law enforcement were described as potentially "massive" (Interview 08 2000, pp. 370–373):

> [according to t]he way the [RIP] Bill is written at the moment, law enforcement—which could include [. . .] minor officials—could actually ask for the most obscure types of data, which [. . .] service providers are currently gathering. And there is nothing in there that says they couldn't do that. So, they could come along and say 'we'd like to [. . .] have a list of all the people with a 'z' in their name, who emailed all the people with an 'x' in their name last Thursday afternoon'. And that'd just be [. . .] communications data. The [. . .] service provider would say, 'well, for heaven's sake, I don't collect that, you know, I don't know that!'. And they'll say 'well, I'm sorry it says [in] section 42' or whatever 'of the Bill [that] I can ask for anything I want and you've got to give it to me'. And they'll just sort of stand there until something happens. [The] service provider would have to comply [Law Enforcement Liaison, large British ISP association, Interview 29 2001, pp. 421–431; see also Interview 08 2000, pp. 358–362, 374–378, Stewart 1999; de Stempel 2001].

There is no evidence to suggest that the wealthier, more profitable telecommunications corporations were in any major way concerned about the logistical costs associated with the above prospects (Interview 04 2000, pp. 299–301; Interview 17 2000, pp. 71–75; Interview 12 2000, pp. 525–532; Interview 07 2000, pp. 11–16; Interview 01 2000, pp. 466–468; HCTISC 2000), which they described as "marginal" (Interview 07 2000, p. 10):

> British Telecom, being a very large operator, [. . .] are already carrying probably too much machinery to deal with [communications interception]. And [. . .] there's probably room for efficiencies [Security Officer, large British TSP, Interview 07 2000, pp. 33–35].

That statement reflected the size and funding of British Telecom's Security Department, which in the mid-1990s was particularly sizeable and gave an idea of the scope of the company's security and investigation budget. The Department reportedly consisted of no less than five different sections: Directorate of Security and Investigation; Commercial Security Unit; Specialist Services Unit; Computer Emergency Response Team; and Investigation and Detection. The latter of these, which was based in Milton Keynes, employed more than 50 detectives from a variety of backgrounds, including police, military intelligence, Customs, Inland Revenue and the Department of Social Security (Bleep 1992; Anonymous 1994).

Among smaller carriers, however, there were real fears that the imposition of such logistical costs on them by communications-interception legislation would gradually turn them into a "law enforcement resource" (Bourne 2001), thus rendering them increasingly unprofitable:

> [s]maller [carrier] businesses are concerned about precisely what will be required of them [by RIPA]. For some, the requirement to invest in 'reasonable' interception facilities could mean the difference between profit and loss in the critical early stages of business operation [Anonymous 1999e; see also Interview 11 2000, pp. 83–87, 100–106; Interview 07 2000, pp. 32–45, 61–62; Perry 1999; Hall et al. 1999];

> [large carriers] have such tremendous resources, whereas smaller companies don't have the resources [. . .]. If you take [CALEA's requirements], those are things far beyond the capabilities of the small companies to [. . .] deal with [. . .]. All of the networking and the aggregation of [. . .] the [. . .] data redistribution and the single point [of contact] that's just [. . .] a level way above what these small companies that we represent would ever be capable of doing. So, [. . .] there's surely a problem with small companies having this imposed on them for [. . .] no reason [Technical Director, national association representing over 500 smaller TSPs, Interview 17 2000, pp. 77–84; see also Interview 18 2000, pp. 154–157; McHugh and Cordwell 2000; Anonymous 1997a].

During the IOCA/RIPA negotiations in the United Kingdom, ISPs emerged as an exceptionally vocal segment of the smaller-sized service provider community (Lansman 2000). The reasons for that had primarily to do with the fact the vast majority of the more than 300 ISPs who offered services in the United Kingdom were operating on minuscule budges, a phenomenon that had given rise to what industry insiders often referred to as the "mom-and-pop" ISPs:

> [RIPA regulations are] OK for us, because we run a 24-hour-7 shift [...]. But virtual ISPs have been known to use auto-pilot and go to holidays to, say, the Mediterranean for two weeks, and [there's usually] only that one [person] that has got access to the [customers'] home phone numbers and the home addresses. [S]o, let's say [that] the police know that somebody dialed in from that number and the [. . .] virtual hosting ISP says 'well, [. . .] that's a StirlingNet customer'. Who the hell's StirlingNet? StirlingNet [finally ends up being] a bloke living in some council estate in Stirling. So, [the police] try to get onto them. The phone number's ringing out; [they] send the cop car around. The neighbor says he's on holiday. Dead end. Does the guy go to jail for not responding properly? That's the worry of RIPA. These people are lucky if they have a sales office. Everything [is] outsourced [Chief Executive Officer, large British ISP, Interview 08 2000, pp. 57, 110–121].

Independent cost studies of RIPA in the United Kingdom estimated that, on average, ISPs would be required to facilitate approximately £20,000 to install RIPA-compliant interception technology and £10,000 per year for maintenance and logistical costs (Clayton et al. 2000). Demon Internet, one of the nation's largest

ISPs, calculated its RIPA-related costs to an amount equal to 15 per cent of its total revenue (Sutter 2001). ISP representatives remarked that, if these estimations were indeed accurate, most of the United Kingdom's ISPs would be forced out of business (Interview 01 2000, pp. 199–203; Snape 2001; Clayton et al. 2000; Anonymous 2000a).

6.16 Challenging RIPA: The Illusion of Negotiations

Considering the degree of fragmentation within the United Kingdom telecommunications industry, it should perhaps not appear surprising that the latter's concerns over RIPA appear to have had a negligible impact on the Act's stipulations (Interview 01 2000, pp. 290–291; Clayton et al. 2000). The overall feeling among United Kingdom industry interviewees was that they were unable to leave any considerable mark in the legislation. According to one external observer, who later advised Microsoft on information security,

> certain leading ISPs have clearly tried to [. . .] work constructively on the RIPA problem. But, I think [. . .] that, to a large extent, their overtures have not been appreciated and, to some extent, rebuffed by the Home Office. I mean, it was only, really, [. . .] in [the] middle of last year that there was any contact that I'm aware of—significant technical contact— between the ISP industry and the Home Office. That's extraordinary. Because, clearly, this debate about encryption and the interception of communications in general has been running [. . .] for four or five years. It's simply extraordinary that up until then there was no direct contact between the responsible department [or] the responsible officials and the industry. But that is so (Bowden 2000, pp. 571–580).

A security officer with one of the United Kingdom's largest TSPs at the time, noted that,

> from a Security Department point of view the consultation that [. . .] we had was adequate to tell [us] what was going to happen but there was no question of asking [us] what was going to happen [. . .]. We're not making the law here—the government is. And they, you know, it's handed down to strategy to be made into tactics. [It's v]ery much top down, really. I mean, [. . .] I've [. . .] worked for [name of company] since 1969 and [. . .] I've seen governments come along [. . .] and hunk lumps of [legislation]. [S]o it's not a question of [whether] you [. . .] trust them; it's a question of you just accept[ing] what they do [Technical Specialist, large British TSP, Interview 25 2001, pp. 52–58, 73–76; see also Interview 10 2000, pp. 117–120, 130–132].

Additionally, United Kingdom TSPs were of the opinion that there was little genuine desire for fruitful negotiations behind the British government's call for the IOCA/RIPA consultation. Rather, it was believed that the government was keen on creating the illusion of negotiations between industry and law enforcement, in an attempt to prevent over-zealous reactions by concerned TSPs:

> I mean, it doesn't really matter whether we trust the government or not, because we weren't [. . .] really negotiating with them [Interview 25 2001, pp. 68–70];

the industry felt that this was being used as a sort of political selling job, to reassure Parliament at this time that, 'yes, industry is working very closely with us, we've got a partnership' and all this lovely rhetoric. But, in fact, the ISP industry is not feeling particularly well consulted or particularly frequently consulted or, when it is consulted, that its message is being listened to [Bowden 2000, pp. 591–595];

I believe they have already drafted the Bill. I believe the Bill was drafted and almost signed, sealed and delivered before someone had this afterthought saying 'hang on, we've got to have a consultation process here' [...]. One has the feeling that it's not out of concern that the Bill might not meet our requirements. It's more out of concern that we would make a lot of noise against the Bill. And it's a certain amount of. . .'it is better', I suppose, 'to have them on side where we can see and possibly influence what their reaction is going to be as opposed to not have them on side and then suddenly hit them with it and open the floodgates off'. But I suspect the general view abroad is that we don't know how much of it is genuine consultation or how much of that is spin and marketing on the part of the government [Regulation Officer, large ISP association, Interview 10 2000, pp. 262–264].

In short, British industry interviewees displayed a consistently elevated degree of skepticism about the ability of the private telecommunications sector to steer the legislation in a more industry-friendly direction. In their view, the government was far more open to the views of law enforcement and intelligence agencies than of the private sector.

6.17   Challenging RIPA: The Fraud Card

Interviews unearthed a particularly interesting element in the RIPA negotiations between industry and law enforcement in the United Kingdom: namely, many of the carriers who were consulted in the debate requested—and could well have been promised, although no proof of that has ever emerged—that they receive increased assistance from law enforcement in combating telecommunications fraud in return for adopting a cooperative stance over RIPA. In the run-up to 2000, telecommunications fraud, ranging from old-school phone-phreaking to cellular phone-cloning, and from network hacking to email spamming, had grown dramatically, costing the wireless and wireline industry in excess of £400 million annually (Interview 09 2000, pp. 322–324). Indeed, throughout the 1990s, telecommunications fraud was the British industry's topmost financial and logistical problem (Roberts 1996). Admittedly, one of the reasons behind that dramatic increase had been the failure of British law enforcement agencies to respond to telecommunications fraud with the same investigative urgency with which they tackled traditional criminal activity. Tim Pearson, Chairman of the United Kingdom's ISPA, said in 1999:

I know from my own firm's experience [that] the situation [at the moment] is that one tends to try and get the crime reported in an area when one notes that there are at least one or two policemen that one might be able to have a conversation with who might understand the issue. We have been to local police stations to report a crime on the Internet and you do not get very far. By the time you have explained routing and domain names, the policeman has other things to do, obviously [HCTISC 1999a; see also Interview 08 2000, pp. 270–279; Interview 09 2000, pp. 181–184]

In interviews, fraud control managers at a large British wireless carrier expressed their frustration with the absence of adequate governmental response to telecommunications fraud:

> you're dealing with a [...] government instrumentality [and] with [...] limited resource and a massive remit. And so [...], I mean [to a police officer's ears] it sounds like trying to convince [them] to come out and get a little old lady's cat down out of a tree [Fraud Control Manager, large British TSP, Interview 06 2000, pp. 180–184].

> I think there's been little movement in that direction—to get the police to understand our perspective. [T]here is a long, long way to go. If [...] we have a customer that is affected by a [telecommunications fraud] crime, we expect them to be able to go to their [...] local police station and report it and be taken seriously. And not to be rebuffed because [the police] don't understand it. Most of the times we have to talk our customers through a step of events that they need to do in order to be taken seriously [Fraud Specialist, large British telephone service provider, Interview 24 2001, pp. 44–49].

The appearance of the RIP Bill, therefore, was seen by segments of the United Kingdom industry as an opportunity to push for the elevation of the status of telecommunications fraud in the eyes of law enforcement, in return for a co-operative attitude in law enforcement communications interception requests. In a 2000 presentation before a large group of law enforcement liaison officers of United Kingdom wireless and wireline carriers, the United Kingdom telecommunication industry's designated Single Point of Contact under RIPA characteristically described the industry's hopes for a new regime under RIPA:

> [up until now] law enforcement would not necessarily take telecoms crime, or telecoms fraud that seriously. It was almost like the blue collar-type crime—it doesn't affect anybody, nobody's heard, shrug of the shoulders and move on. That attitude is fast diminishing. And certainly now with single points of contact and, more recently, with accredited officers, they know that if they don't co-operate with our investigations on crime and help us they're not going to get the same in return. So, it is a certain amount of give-and-take in that respect (Anonymous 2000b).

Industry interviewees overwhelmingly reflected similar hopes for a change in law enforcement's attitudes over telecommunications fraud in the post-RIPA environment:

> we'd like to think that we're not easily pushed over [by law enforcement]. We would like to be seen [as] having an important role and [...] as a [...] a partner with the police, rather than [them thinking]: 'we can gain as much information as we'd like from you and you can't say no; you've got to just do it'. It [...] doesn't work like that—it's a two-way thing [Fraud Control and Revenue Assurance Manager, large British wireless communications service provider, Interview 04 2000, pp. 37–40];

> in the area I'm working in at the moment we're investigating fraud against [our company]. [A]nd, if we are seen as being adaptable and [...] law enforcement-friendly, when we go to them and say 'will you help us to arrest these culprits?' they won't bring forward bureaucratic or procedural reasons—which they could. They could say 'we have other priorities'— they've always got a murder running somewhere. They could say that they have priorities. They'll help us knowing that [...] we are professional and [...] police-friendly. So there's a bit of push and pull in it [Security Officer, large British TSP, Interview 07 2000, pp. 145–151];

[o]ccasionally we want help from the police and [...], yes, it helps to have a good relationship. If they want some help from us, we give it on the idea that [...] when we want help from them we'll get it [Security Officer, large British ISP, Interview 27 2001, pp. 279–282].

The question of whether there had been any direct or indirect connection between digital communications interception assistance by carriers and priority-designation to telecommunications fraud by law enforcement during CALEA negotiations was asked during all United States interviews. Without exception replies were negative (see for instance Interview 16 2000, pp. 202–205; Interview 17 2000, pp. 192–197; Interview 19 2000, pp. 771–788).

6.18   Challenging CALEA: Climbing the Hill

In the United Kingdom, aggressive competition appeared to have suppressed the effective collectivization of the industry's voice in regards to IOCA and, later, RIPA. In the United States, equally antagonistic market rivalries within the industry sector appear to have been counteracted by a series of very sophisticated lobbying campaigns by the industry's well-paid representatives. Once the industry realized that a CALEA-type legislation was favored by a law enforcement-friendly Congress, its collective representatives concentrated their efforts on preventing law enforcement's access to decision-making processes that concerned technical standardization (Interview 18 2000, pp. 18–20, 20–24). That was eventually achieved, which meant that the telecommunications industry effectively denied American law enforcement and security agencies the leading role in the technical design of digital communications-interception hardware and software. The effectiveness and thoroughness of the industry's lobbying efforts were recognized by both carrier and law enforcement participants in the negotiations on issues concerning CALEA:

[we] organized a lot of [...] activity [...] up there [on Capitol Hill], because [...] the United States Telecom Association, the Organization for the Promotion and Advancement of Small Telecommunications Companies, the associations along with the [...] large [telecommunications] companies have a lot of people [...] who work at the [Capitol] Hill. This [...] is why some of the language [...] got [...] changed—because of a lot of the work. Because, when [CALEA] first came out [...] it was not very good at all. And then we went through the interviewing process, the compromise, etc. [Director, National Services Planning, large TSP association, Interview 37 2000, pp. 81–88; see also Interview 20 2001, pp. 128–130];

what CALEA represented was a compromise between the telecommunication industry, privacy advocates and law enforcement [CALEA Compliance Attorney, large United States TSP USA, Interview 20 2001, pp. 23–26];

representatives like Don Bender and [...] others with [the United States] Telecommunications Industry Association and the Cellular Telecommunications Industry Association [...] have a huge, powerful and influential presence with Congressional staffers, as well as with the FCC. I would say that we are mere blips at our relationship with government in that respect [...]. The industry has very well-paid talent [...], people that contribute to soft money, political action committees, not to mention their regular contributions to the political process [FBI Special Agent, CALEA Implementation Section, Interview 19 2000, pp. 402–406];

I don't think [that] we were really aware of the enormous lobbying power of the telecoms [industry]. I don't think we had [...] prepared for the challenge that we [...] faced. I don't think we could have [prepared], frankly [FBI Special Agent in Charge, Interview 35 2001, pp. 130–132].

Beyond their own ability to orchestrate sophisticated lobbying campaigns, what made the crucial qualitative difference between the legislative power of United States and United Kingdom TSPs on CALEA and RIPA respectively was the fact that the former were actually invited to a genuine round of negotiations by government. Hence, despite the legislators' national security-oriented preconceptions, the carriers were given "the opportunity, as an industry, to be involved" (Interview 17 2000, pp. 125–127) in the debate. By 1998, the United States telecommunication industry's campaign had been effective enough to allow civil liberties observers to assert that "under CALEA, Congress made it clear that privacy interests and industry interests are paramount over law enforcement concerns" (Dempsey and Oram 1998). The underlying meaning of that statement directly contrasted with the impressions of British industry officials, who were involved in the RIPA debate:

I think that [...] if there are suspicious motives [behind RIPA] being driven by, let's say, the Security Services or whatever, then there's not that much chance of getting them changed. They've got a lot more power and influence that any industry body has, so we're kind of [...] nibbling away at the edges [of the legislation] a little bit. If there is some fundamental, underlying [...] suspicious motive behind all of this, then [...] I'm afraid that there isn't much that we can do about it [Law Enforcement Liaison and Fraud Specialist, large British ISP association, Interview 30 2001, pp. 209–216].

6.19 Defending RIPA: Backdoor Lobbying

Clearly, the above quote reveals a sense of being overpowered by political decision networks that were detached from the industry's economic dilemmas, and operated on an exclusive level that was monopolized by national-security priorities and concerns. In some cases, British industry, as well as civil liberties-oriented interviewees, did air the belief that ministerial ears were freely lent to members of the law enforcement and security services, who were often all too keen to remind elected officials of the non-negotiable qualities of law and order:

a party in opposition has all sorts of wild ideas about what it is going to do when it gets into government and [...] the policy appears to do a U-turn within days of getting into power. And you can't help but think that, in fact, it's not really [...] within days of them getting into power, but it's within days of them [...] having access to certain senior civil servants, who just say 'I'm sorry, but this is the way it all works; we do need these powers for these very good reasons and, I'm sorry, we don't really care what your policy was when you were in opposition. You're in government now and you've got to change your tune. You've got to behave like government. You can't behave like the opposition any more'. At which point lots of U-turns in policy get done at this kind of area [Regulation Officer, large British ISP association, Interview 10 2000, pp. 229–231];

from working in this area for a number of years, one certainly gets the impression that if you are a Minister and your security agencies come to you and say 'well, look Minister, of course you can do what you like, but unless you give us these powers, in a couple of years you will have ghastly headlines about [. . .] gangsters running riot all over the country'. And, if you're a Minister and you [. . .] don't really have any detailed knowledge of this area or the technologies or the criminology, you will ask a few questions but, basically, you have to trust what you're being told [Bowden 2000, pp. 241–248];

[i]f you take an average politician to the MI5 or MI6 or Aldershot or Hereford or Cheltenham or any of these places [. . .], they sit in awe: 'ah, there's all sorts of secrets in here; this is where James Bond lives and breathes'. They're just in awe. They're absolutely in awe. They're like kids. They really are. It's pathetic, it really is. And they will just suck it all up. You know, who are we? We're just waiters. . .they don't care [Security Officer, large British ISP, Interview 27 2001, pp. 170–174].

6.20   Defending CALEA: The Rhetoric of Emotion

United States Law enforcement agencies began pressuring for a CALEA-type solution during the late 1980s. They intensified their campaign in 1992, when it became apparent that the industry was not committed to a voluntary scheme of digital communications interception provision (Blair et al. 1995). There is little doubt that federal law enforcement made use of its direct access to the government in an attempt to gain a policy advantage over the industry:

the [. . .] influence [. . .] of [federal law enforcement] agencies is [. . .] funnelled and directed through one or two voices in those agencies [. . .]. And they work quite gently with Congress. They cannot directly lobby Congress, [. . .] by law [. . .], they can only educate Congress. So there's some education that's going on there and [some] discussions [FBI Special Agent and Telecommunications Industry Liaison, CALEA Implementation Section, Interview 40 2000, pp. 408–420];

there were a lot of people in Congress that supported [CALEA] and the FBI was pushing it very hard [. . .]. I would suspect that the Department of Justice lobbied pretty heavily at the back room [CALEA Director, large national TSP association, Interview 39 2000, pp. 17–18, 125–127; see also Interview 17 2000, pp. 149–150];

when you have as many FBI agents as they have and you send a bunch of them to Capitol Hill to do the lobbying. . .we often times found when we were lobbying that the FBI had been in just ahead of us with a bunch of agents and their Director [and] they'd been talking to influential members of Congress. This is why we made a lot of changes to CALEA early on and we couldn't get any later—because of the influence of the FBI [CALEA Compliance Attorney, large United States TSP, Interview 20 2001, pp. 170–175].

It is equally apparent that law enforcement's campaign was based less on technical and operational details and to a much higher extent on emotional rhetoric focused on domestic terrorism targeted at United States interests, and protecting children from paedophiles and other criminals. The example of a case where intercepted communications assisted the FBI in preventing the planned kidnapping and murder of a young child for the purpose of making a 'snuff murder' film was constantly cited by proponents of enhanced digital communications interception capabilities for law enforcement at almost every given opportunity (Denning 1993; Barlow and Denning 1994). Other instances of carefully-constructed rhetoric included a 1994 public relations press conference, in which a law enforcement lobbyist, Detective Brian Kennedy of the Sacramento Sheriff's Department, remarked that what bothered him was that "there could be kids out there who need help badly, but thanks to this encryption, we'll never reach them" (Baker 1994; see also Anderson et al. 1999; Freeh 1997):

> we brought numerous instances, newspaper articles, clippings, sob stories, women who were carjacked and their phone was taken with them and we had to track the phone and get the content. So [. . .], we did a good bit of flag-waving [FBI Special Agent, CALEA Implementation Section, Interview 19 2000, pp. 179–183].

There is evidence to suggest that, although this type of campaign was rather effective in recruiting elected representatives in favor of CALEA, it did not have the same impact on industry executives, who were primarily concerned with the legislation's economic impact on their businesses:

> I mean, every time we would go out to lobby what we would hear on [. . .] Capitol Hill was 'you're going to take away the ability to go rescue people who are being kidnapped' or whatever. And [. . .] law enforcement uses that crime scenario very effectively [CALEA Compliance Attorney, large United States TSP, Interview 20 2001, pp. 167–170];

> I would say it had little effect on [the carriers]. And, after a while, we dropped that as a *modus operandi* because it just was water off a duck's back [FBI Special Agent and Telecommunications Industry Liaison, CALEA Implementation Section, Interview 40 2000, pp. 181–183];

> politicians [were interested in] not appearing soft on crime and criminals, [whereas] the industry people were: 'show me the money!' [. . .]. So we had to approach them in a different manner [Police Officer with experience in undercover and wiretapping operations, Interview 14 2001, pp. 139–141; see also Interview 17 2000, pp. 150–152].

Ultimately, law enforcement's insistence on emotional rhetoric during lobbying campaigns might have helped them have CALEA approved by a Democratic administration, which was initially viewed by some as adversarial to law enforcement's interests (Interview 19 2000, pp. 374–375)—although external observers remarked that the United States President at that time, Bill Clinton, had been a supporter of communications interception since his days as Governor of the State of Arkansas (Willing 2000). But that same approach had little effect on the detail of the legislation, which was primarily controlled by the industry:

> our [campaigning] efforts were effective in that [. . .] a CALEA sort of plan was approved. But we [. . .] didn't get what we wanted [. . .]; we were not happy with most of the important provisions [. . .]. We wanted CALEA features implemented everywhere and [. . .] as soon as practically possible and that clearly didn't happen [Police Officer with experience in undercover and wiretapping operations, Interview 14 2001, pp. 452–455, 457];

> [CALEA's] first drafts were certainly more law enforcement-slanted than what came out of the [negotiations] [FBI Special Agent, CALEA Implementation Section, Interview 19 2000, pp. 380–381].

6.21   Citizen Participation: The Crucial Absence

In an important sense, the introduction of CALEA and RIPA marked one of history's rare instances where the communications interception debate spilled over into the public arena, with numerous articles on the subject appearing in the popular press. However, the negotiations that gave shape to the two pieces of legislation were largely, if not solely, monopolized by two multifaceted sets of actors: law enforcement and the telecommunications industry. In the United Kingdom there were attempts by citizen-led civil-liberties organizations to enter the upper echelons of the negotiations process. Although similar attempts in the past achieved a modest degree of success in terms of affecting legislation related to electronic communications (Interview 01 2000, pp. 142–145, HCTISC 1999a), RIPA was kept almost entirely away from the policy reach of such groups. Professor Akdeniz and Caspar Bowden, who represented part of the British civil-liberties lobby, stated:

> in all these meetings and [. . .] regulatory initiatives no one is taking into account individual rights and liberties. The industry is concerned about their financial interests [and] law enforcement wants access to this and that [. . .]. So it is very difficult to establish a dialogue from the civil liberties perspective, because [we are] not IBM or even Demon Internet. Well, when you're Demon Internet or [America On Line] they feel like they have to listen to you. But [. . .] who do we represent? [Akdeniz 2000, pp. 210–214, 259–266];

> the reality is that, [. . .] for as long as I have been involved in this [. . .], civil liberties representations are not formally recognized at all by the government. If you go back through the literature of press statements, the twists and turns of policies over the past three or four years, I don't think you will find one official acknowledgement of any civil liberties concern in any of this area—and one saw this through the debates on key escrow, as well [Bowden 2000, pp. 192–197].

In order to assume a position of relative influence, civil liberties groups attempted to approach the industry and introduce a civil liberties angle to the latter's rhetoric— a formula that had been successfully implemented during an earlier debate on key escrow (Akdeniz 2000, pp. 182–183; see also Bowden 2000, pp. 315–322). Indeed, it was that effort that led to the creation of the Internet Users' Privacy Forum (IUPF), an industry/civil liberties scheme that was not emulated in the United States (see Cole 1999a, b). However, in approaching industry in this manner, civil liberties groups were somewhat forced to tone down their critical attitude against the RIPA, and eventually found themselves involved in the—not always wholesome—process of transforming their inherently political demands into consumer demands. According to Bowden,

> it is only industry lobbying that seems to be effective in changing government policy in this area [. . .]. It is [. . .] true that, if industry perceives that [a] large amount of civil liberties is concerned, then that tends to encourage the industry to be more forthright in their representations to government [. . .] because, [. . .] if the [industry] feels that they're [. . .] being asked to operate a regime which [will] not be acceptable to [. . .] customers in the market place, then that worries them on strictly commercial grounds (Bowden 2000, pp. 205–211).

In addition, there is little evidence to suggest that the paramount industry actors paid any significant interest to the criticisms and propositions of the civil liberties community. In fact, the relationship between civil liberties and industry groups in regards to RIPA often appeared adversarial. Few industry interviewees were prepared to condemn the civil liberties community on record; but, with the recorder turned off, many described the latter as "mosquitoes"—in the sense of annoying legislators while having little long-term effect—or "extremists':

> I guess for, say, the likes of Caspar Bowden and . . . and the people who are interested in the . . . Liberty and organizations like that, they . . . I mean . . . they have a different view. But [as far as we are concerned] the oversight process is there, in the [. . .] last part of the [RIP] Bill [. . .] and there is some [. . .] recourse there. And [we] don't agree [with] all the people that are concerned about issues of privacy and personal security and [. . .] going to an extreme [Head, Regulation Affairs Department, large British TSP, Interview 32 2001, pp. 320–326].

On the basis of the information provided for this study by civil liberties advocates (Sehgal 2000; Cohn 2000; Courtney 2000), it can be asserted that, although the impact of the civil liberties community on CALEA was not as powerful as ideally expected, civil liberties advocates were not marginalized from the legislative debate to the extent that their United Kingdom colleagues were:

> [w]e were invited to participate in negotiations in late 1993, so we did have an opportunity to contribute to [CALEA] prior to its enactment [which was] in October 4, 1994 [Sehgal 2000];

> either through [members of] Congress or through the [telecommunications] industry, we were able to score a number of significant victories in [the] CALEA [debate], particularly with roving wiretaps and the [J-STD-025 or] J-Standard [Cohn 2000; see also Courtney 2000].

The relative impact of civil liberties on the legislation was also acknowledged by one law enforcement interviewee and even by industry participants in the debate:

> [the industry] took some hits from [. . .] the privacy groups, which w[ere] the Center for Democracy and Technology, as well as [the] Electronic Frontier Foundation. They too had input into the process and there was some give-and-take there [FBI Special Agent and Telecommunications Industry Liaison, CALEA Implementation Section, Interview 40 2000, pp. 367–370];

> ah, they [the civil-liberties groups] are very influential. Yes. Yes. The interesting thing about the [. . .] privacy sort of civil-rights [issue], is that it's sort of a non-partisan issue and there you'll find [. . .] their members being from both the Democratic and Republican party that are very pro-privacy [. . .]. So, it doesn't necessarily cut off on party lines [CALEA Director, large national TSP association, Interview 39 2000, pp. 115–121; see also Interview 17 2000, pp. 133–137; Interview 16 2000, pp. 131–135; Interview 20 2001, pp. 368–370; Auble and Bender 2000].

Broadly-speaking, however, neither the United Kingdom nor the United States digital communications interception debate was particularly enriched by the adequate presence of grass roots, citizen-based organizations. There are reasons to believe that the void was filled up by generalized assumptions about public attitudes held by the participants in the debate, which may or may not reflect reality and which may or may not have influenced the spirit of the legislation:

> I think that the vast majority of the public—and, after all, most of the customers of ISPs are the public—probably prefer society to be a safer place because of law enforcement's

activities and therefore would say 'well, of course you've got to go out and catch the criminals. I don't want my computer hacked, so please [act]. I don't want people sending me viruses, [so] please go out and catch these people and stop it happening to me' [...]. Therefore, I think, the public at large [...] is very pro-law enforcement. [I]f there was a referendum they'd bring back hanging, all right? And a lot of those libertarians forget this [Regulation Officer, large British ISP association, Interview 10 2000, pp. 614–622];

the public cannot be trusted to make a reasonable decision about [communications interception]. I think the public, when it comes to issues of cyber-crime, will definitely err on the side of [...] law enforcement and say 'yes, for heaven's sake, can't they catch these people? Here, can I lend you my gun?' [Interview 10 2000, pp. 627–630];

you know, no, I don't [. . .] trust the general public and that's not just my [personal] opinion [. . .]. The general public usually—nine times out of ten—is incapable of handling the truth in these matters. Look at, for example, what happened with the death of Princess Diana in England. I mean, they just go mad. So, in my opinion, it makes absolute sense for some of these issues to remain confined to an executive [. . .] governmental level [Sergeant, large United States police department, Interview 15 2000, pp. 631–637].

Reflective of these attitudes is a comment made by former United Kingdom Minister of State Jack Straw in Parliament in 2000:

I also explain to constituents at my surgeries that, given the weight of the threat from international crime and terrorism, the prospect is remote that we would devote resources to intercepting their telephone calls. However, I then provide no reassurance when I say that I cannot tell them whether their telephone is the subject of intercept because otherwise others who might be the subject of intercept could come along on fishing expeditions. It is logically a difficult position to explain to individuals, and it is difficult for people to understand that (House of Commons 2000).

It is admittedly rare for negotiations on a topic as politically sensitive as communications interception to be visible in the public domain. The public attention given to CALEA and RIPA was unprecedented. Nevertheless, the negotiations between the two main actors, the telecommunications industry and law enforcement, were completely shielded from public view and remain so to this day.

6.22 Communications Interception and Accountability

This confinement of communications interception-related information to executive levels has historically sparked a number of popular and scholarly debates about the degree of accountability, or lack thereof, that characterizes state-sponsored communications interception practices (Fitzgerald and Leopold 1987; Theoharis 1971). The fact that there is little documented information about illegal communications interception by law enforcement agencies in the United States and the United Kingdom has not prevented civil rights watchdogs from being suspicious:

as to what's been done in the past, [. . .] most of what one hears [. . .] is anecdotes. But there certainly are plenty of anecdotes [. . .] right up to the present day, about informal approaches being made to telecommunications companies and access being granted [. . .] on a sort of first-name-term 'matey' relationship basis, both to the communications data and to content without a warranted procedure being in place [Bowden 2000, pp. 467–472];

[i]f you build an infrastructure that encourages wiretapping, it will be abused. The only question is: how many years or decades will it take to find out it's been abused? [John Gilmore, Electronic Frontier Foundation Co-Founder, quoted in Poulsen 1999].

In the United Kingdom, the closest government sources come to admitting instances of illegal wiretapping is by reporting for the public record a number of instances where communications interception operations have gone wrong. Such instances are published in the Interception Commissioner's annual report to Parliament. The following characteristic quote is from the latest such report:

[in one] case the Security Service applied for a revalidation of a warrant, but the submission listed an incorrect operator. This initial error was not identified by the Home Office and the warrant was renewed including the incorrect operator. This renewal was therefore invalid. This error only came to light when an approved request to modify the warrant to delete one number and add another was submitted to the original operator. The error was compounded by the fact, confirmed by subsequent investigation, that given the particular circumstances of this operation, the intercept had not been suspended when it should have been. This increased the period of unlawful interception. All product from the total period subject to the error has been destroyed. The error reflects badly on all parties concerned, the Security Service, the Home Office and the operators involved [Command 4778 2000, Sect. 45].

Such incidents—which are not few—are usually attributed to erroneous judgements and logistical shortcomings. But United Kingdom interviewees painted a rather different picture, namely one where illegal communications interception and data acquisition by law enforcement are often the outcome of calculated intent, rather than human error:

[t]hings do go wrong. In the past, the typical things that have gone wrong have been like, sort of, a police officer who suspects his neighbor of having an affair with his wife and [...] wanted to find out if she'd phoned the mobile or something like that [Technical Specialist, large British TSP, Interview 25 2001, pp. 107–109; see also Interview 08 2000, pp. 298–305];

I wouldn't [...] say that there isn't any [illegal communications interception] going on, because I would be lying. It does happen, though I know it's rare nowadays. It depends on the arrangements with local constabularies. When I started working for [name of company], I mean, there would be times when officers would [call] the technicians [...] and request information, rather than going through [the] local branch managers. I mean, because they would eventually get to know each other and so [...] that's what would happen [Law Enforcement Liaison, large British TSP, Interview 23 2001, pp. 401–408; see also Parr 1999; HCTISC 1999b].

Another phenomenon, which usually occurs with the larger, more established telecommunications companies, is that they will often make considerable effort to honor law enforcement requests, despite the latter's lack of proper authorization, and while remaining within the spirit of the law:

I'll give you an example: somebody says to [us]: 'we want an ex-directory number' and we don't want to give them that ex-directory number unless we have to. So [...], then we say: 'why do you want it?'. And they say: 'we want to telephone this person'. Then we say: 'fine, we'll set the call up for you and you can speak to them, but we won't give you the number'.

Then we avoid having to do the rigmarole for giving them permission for an ex-directory number. So we [...] don't have to give them the number, but they get what they want. And this is [...] I mean, [such] examples [occur] very often, I would imagine [...] certainly on my team, in the past, when we've handled it [...]. We would say to them 'you can't have that but you can have this and it solves your problem'. And so [...] we wouldn't refuse. I don't think we would ever refuse [...] a request arbitrarily [Security Officer, large British TSP, Interview 07 2000, pp. 127–138].

It is worth noting that, on the whole, downright refusals of law enforcement communications interception requests are very rare on both sides of the Atlantic (for the United Kingdom, this was confirmed by interviewees in Interview 04 2000, pp. 154, 156–158; Interview 06 2000, pp. 96–98; Interview 07 2000, pp. 117–118; Interview 08 2000, pp. 264–265; Interview 09 2000, pp. 260–261; Interview 10 2000, pp. 289–290; Interview 11 2000, pp. 474–477). According to one FBI interviewee, there had been only one case in living memory when a carrier refused to honor a request (Interview 19 2000, pp. 573–592). But other FBI sources have revealed that

there have been anecdotal reports of instances where carriers have refused to provide assistance to Law Enforcement even after being presented with a facially valid court order in circumstances where carrier personnel did not recognize a particular judge's signature or where the description of the carrier service to be included in the intercept did not precisely match the carrier's brand name for that service (Morris 1998; see also Dempsey and Oram 1998).

The relevant scholarship agrees, however, that the last such incident prior to the enactment of CALEA occurred in 1968 (Boucher and Edwards 1994). One MI5 officer explained that most unauthorized law enforcement-sponsored communications interception concerns the acquisition of communications data, rather than content, because it is easier to obtain, and can be facilitated by the carrier without the requirement of much logistical support. This practice, which is known in industry and law enforcement circles as a "fishing expedition', is quite straightforward and can involve a large number of telephone subscribers:

[i]n the past it [...] has not been unknown for law enforcement agencies to 'fish'. [S]ay they put in a suspect and on the suspect there are six telephone numbers and for some reason they relate this to the crime in hand. So, they ask for the subscriber details on the six telephone numbers and they are provided.

Q: OK, so that's a fishing expedition.

A: No, it's not. The fishing operation hasn't even started yet. They get the six numbers and then come and say 'we want the billing record for the last 3 months for these six people. That's the fishing operation. And they should then get the billing records and they go from the billing records to see who their six people have called. And then you say 'right, let's analyze those; and let's get the billing records of those 500 people'; and you can see how you can build that up. Now, with the modern analytical tools you can put all that information in and you can get a picture out. Right. But that's a fishing expedition—because, at the end of the day, whilst it would be quite genuine for the first six [people], no way is it genuine for the six billing records, over whatever period, for those six [people]. And you then start to move away from what is immediately involved [Officer, MI5, Interview 12 2000, pp. 409–425].

However, the interviewee—who had experience in working for both MI5 and MI6—went on to assert that this practice had been gradually fading out ever since British Telecom's privatization and the introduction of a number of legislative safeguards:

I know of instances where, two years ago, 'A' would ring 'B' and say 'who is on this telephone number?' and B would say 'no problem Mr. Smith'. [T]here was no accountability. No one in the past has had to account for [. . .] anything. You know, I mean, [. . .] we've only got to go back, I suppose, ten [or] fifteen years, that if a policeman said to you 'I want to know about ABC', because he was a policeman, you'd tell him. I mean, you know, he was the policeman, it was his job, he was investigating so you'd tell him [. . .]. In terms of the industry, [there have been] arrangements with law enforcement, which have not necessarily been as tight as they should have been. But that now is a thing of the past. I mean, now no one gets any information, personal details of any customer or any subscriber, unless there is a clear order to put trail on where that information is going and why it is required. I mean [. . .], I have seen [it in] recent cases; I mean, that's a sea of change in the last few years, brought about mainly by [the 1998 Data Protection Act or] DPA and the requirements for DPA [and] re-enforced by the changes in the Telecoms Act, in [the] Freedom of Information [Act], in [the] European [Convention on] Human Rights, [which] has been proposed in the United Kingdom legislation, in terms of human rights [. . .]. In the past, the fact that the documentation didn't follow through was neither here nor there. Now it has to. And if it's not, then within 24 hours the company is back to that Police Force in question and wants to know why [. . .]. If they rung 'B' today, 'B' would say 'DPA form'—end of story. And it [has] happened in recent past almost on a daily basis [and] it's getting to the point now where 'A' wouldn't even ring 'B' unless there was a DPA form—end of story. So there has been a much tighter control in that level of information—that lower level of information [. . .]. I wouldn't hand-on-heart say 'it would never happen'. Because, you know, at the end of the day, police are police, they're human beings and [. . .] will try [. . .] it on sometimes—no doubt. But they will get short shrift to where before they got almost [. . .] blind obedience, because [. . .], at the end of the day, they were the good guys, they were asking and were given it [Interview 12 2000, pp. 327–395, 430–435].

In a presentation before a large group of law enforcement liaison officers of United Kingdom wireless and wireline carriers in 2000, the United Kingdom telecommunication's industry's designated Single Point of Contact under RIPA said the following about fishing expeditions:

[w]e've all at some time or another experienced a policeman ringing us up at some ridiculous hour saying: 'this is urgent, this has got to be done' only to find at some later date that was not the case. Hopefully, fishing expeditions are a thing of the past—and we've all had them (Anonymous 2000b).

There were strong indications, however, that warrantless government-sponsored communications interception requests were occurring increasingly in the ISP industry, where both law enforcement and carriers were unsure as to the legal requirements that had to be present before communications interception could be facilitated (HCTISC 1999a; Interview 01 2000, pp. 307–309; Interview 05 2000, pp. 454–457). Consequently both sides—the government agencies and the service providers— usually chose to opt out of the warrant process completely:

Q: [Y]ou don't have to answer this question, but, according to your knowledge, have communications service providers—ISPs in our case—always been served with warrants prior to assisting law enforcement with [. . .] interception of communications?

A: I haven't had a single warrant. In all the occasions the cops were completely flatfooted. They had no [. . .] idea what they were doing. What they were looking from us was whatever they could get [. . .]. They came out here and they said: 'we'll get you a warrant' and . . . whatever. They said: 'the reason we haven't brought one is we're not quite sure what it is we're applying for'. They said: 'can you show us what you've got and what you can do?' [Chief Operating Officer, large British ISP, Interview 26 2001, pp. 203–211, 219–222];

[t]he smaller ISPs are generally the worst in protecting your confidentiality. They 'roll over' easily. I've called and they've told me the guy's real name over the phone without a subpoena or anything' (Feldman 2000);

[a]t present, the police regularly make us aware of a serious crime, which they cannot investigate without us providing contact information for one of our customers. They know the Internet identity of the person they wish to investigate but only we, as their ISP, can link this to a real world identity. We are able to supply this personal information [...]. We believe that this arrangement (famously described by Peter Sommer as 'cozy'), which does, of course, allow the police investigations to proceed when they would otherwise be stymied, will break down in the face of a specific commitment to our customers that their personal details are confidential. In the relatively near future ISPs are likely to insist that the police always have a warrant before data can be made available to them [Hall et al. 1999].

In the run-up to the introduction of RIPA, ISP representatives had also complained to the British Parliament that, on numerous occasions, police had required them to "check their records just to see if a suspected criminal might possibly have an e-mail address" (Delio 2001). Rachel Basger, regulatory manager of World Online, said in 2001 that that she had been

asked [by law enforcement] to provide a list of users who had a specific zip code shared by someone who was under investigation, just in case the suspect had an account with World Online. Perhaps I got the short straw and pulled the only really stupid request. But from what I hear, I worry that it was not an isolated incident, Basger said (Delio 2001)

In the United States, instances of unauthorized state-sponsored communications interception were equally common in the 1990s, though industry and government sources were usually unwilling to discuss them publicly. In 1998, for example, a Los Angeles deputy public defender discovered that officers in the Los Angeles Police Department had concealed the role that unauthorized communications interception operations had played in as many as 425 prosecution cases since 1993. The content of telephone calls that had been acquired through the illegal operations had been used to prosecute and convict defendants, but it was never disclosed to the court that they were obtained through communications-interception methods. In the same case it was revealed that police officers had manipulated public communications-interception statistics by using a single court order to obtain multiple taps on many individuals. It was reported that

[o]ne order, facilitated with the help of a cellular telephone carrier, lasted two years and involved the wiretapping of 250 phones, but was reported to state and federal authorities as a single surveillance [Poulsen 1999].

One interviewer, a Sergeant at a large United States police department with significant experience in communications interception operations, described how law enforcement were able to evade effective public oversight in communications interception investigations:

> say you're [...] Police Detective Doe—right?—investigating a car thief, right? And you [...] apply for an [...] intercept [warrant] and you get it, so you start intercepting the car thief's [telephone] calls, right? Well, [...] then you find out that the car thief has been talking to a drug pusher on his phone. But you can't move in to arrest the drug pusher, [be]cause you don't have a probable cause. So, what do you do? You [...] hand over your [intercepted data] to your fellow Police Detective Roger, who then goes ahead and uses it to gather separate evidence and then arrest the pusher—see? So [...] that's what they call a *hand-off technique* [Detective, large United States police department, Interview 36 2000, pp. 189–198].

A number of interviewees also disclosed that, historically, a considerable percentage of law enforcement-sponsored communications interception operations that occur in small-town and rural areas of the United States had been unauthorized, and usually facilitated through the informal relationships formed between law enforcement and carriers in small-town settings:

> you [can have the] sheriff coming in [to the local carrier's offices] saying 'hey, so and so is squabbling with his wife, can you [...] listen in?'—you know. But [...] small types of things like that aren't Title IIIs and it's [...] been done for years in [...] small [telecommunications] companies. They all understand. It's alright. They're good [...] citizens and it's their [...] obligation to do so [Technical Director, national association representing over 500 smaller TSPs, Interview 17 2000, pp. 55–62];

> [b]efore [the AT&T breakup] we just called them [to request assistance]. When I called over to their center and said who I was, they even knew our operators. They said 'oh, yeah, what's the trouble, [his first name]?'. We had a policy that anytime we had a new employee or any time they had a new employee at their phone center we sent our people over there to [...] meet the people, to see the equipment, to see what they were doing. When they had a new employee they'd come over here and [...] and we did the same thing. They would come over, we would send our person and they would meet the new [...] employees [...]. And then those people would send their people back. So we tried to—as much as we could—not to let it go more than 90 days before a new employee on either side coming and meeting one another. So that when ['A'] calls ['B'], he knows that ['B'] is a little fat, short, bald-headed guy and he knows where he's at, he knows what his equipment is and a little bit of his personality, [be]cause he sat and talked to him some [Technical Unit Chief, Emergency Communications Unit, state government agency, Interview 13 2000, pp. 153–164].

6.23 Summary

Digital communications interception legislation was—and remains—a contentious issue, partly because of the dependence that United Kingdom and United States law enforcement and security agencies have on the practice as an investigative technique. During the negotiations over RIPA and CALEA, that crucial dependence was recognized by the carriers, who were either compelled by legislation or by corporate citizenship principles to assist law enforcement.

The industry was generally split on whether RIPA and CALEA were technically achievable. Nevertheless, RIPA and CALEA were partially introduced to allow law enforcement to bypass its lack of technological knowledge by rendering industry professionals responsible for carrying out sophisticated digital communications-interception operations. There were widespread concerns that, in the fully implemented RIPA and CALEA environments, industry engineers would be conducting 100% of communications interception investigations. That could have the effect of making carriers more engaged in questioning the legal justifications of communications-interception operations. It would also place carriers at the center of the battle for power between groups aiming to increase their communications-interception capabilities and groups aiming to augment their communications security. That worried both the carriers and law enforcement, who feared losing control over the micro-mechanics of communications interception under RIPA and CALEA. By 2000, however, the two groups, were gradually being drawn closer by the legislation's operating model.

Another reason why RIPA and CALEA were introduced was to tackle the large number of actors that were part of the mushrooming telecommunications sectors in the United States and the United Kingdom. That was a direct outcome of the sector's deregulation, which took place in the 1980s. It produced both financial and cultural shifts, which drew industry further away from its once intimate relationship with law enforcement.

The negotiations between law enforcement and industry over RIPA and CALEA were largely held in secret and few details are known about them to this day. Neither of the two sides were unified in their interests, which had a negative impact on the industry's ability to shape the legislation, especially in the United Kingdom. One of the few negotiating cards that the British industry was able to use was its offer to be cooperative under RIPA in exchange for more attention paid by law enforcement to growing phenomenon of telecommunications fraud. The American industry had higher success in negotiations over CALEA, and so did the American civil liberties lobby, which worked in cooperation with industry. In contrast, the British civil liberties lobby was excluded from negotiations over RIPA. The implications of the findings presented in this chapter, in light of the events that have unfolded since 2000, will now be discussed in the following chapters.

Chapter 7
The Trajectories of Communications Interception

This chapter examines some of the findings in the qualitative data that were outlined earlier, within the wider scholarly literature on information security and communications interception. Following a comparison between the British and American communications-interception debates, there will be a discussion of the impact of market deregulation on state-sponsored surveillance structures on both sides of the Atlantic. In conclusion, there will be an analysis of the ongoing technical process of integrating communications interception mandates into the very structure of the telecommunications network infrastructure, which is stipulated by CALEA and RIPA. The analysis will also take into consideration the political and cultural developments that are shaping the evolving legal framework of state-sponsored surveillance in our century.

7.1 Comparing Transatlantic Case Studies

This research establishes significant similarities between British and American debates on digital communications interception during the late 1990s. To a large extent, these similarities were caused by the common technical features in the digital telecommunications networks of the United Kingdom and the United States, and gave rise to roughly comparable legislation. The latter aimed to restore state-sponsored communications interception capabilities, which had been taken away by the rise of digitization. Another factor behind the policy parallels found in this research was the broadly similar market framework that characterized British and American telecommunications industries after 1984.

Essentially, the digital innovation of the telecommunications revolution of the 1980s and 1990s black-boxed the relevant technology, and made it impossible for government agencies to access the networks without the direct assistance of TSPs. It was an undeniable reality that elevated the role and importance of carriers in the practice of communications interception. The new interceptions regimes stipulated

© Springer Nature Switzerland AG 2020

J. Fitsanakis, *Redesigning Wiretapping*, History of Information Security, https://doi.org/10.1007/978-3-030-39919-1_7

by CALEA and RIPA recognized that new reality by stipulating the transfer of technical control over wiretapping from governmental agencies to private carriers. This research makes clear that British and American law enforcement and security agencies lamented that loss of control and appeared rather apprehensive about having to hand over the micro-mechanics of communications interception to the private sector. Accordingly, there was a sense of alarm among government officials about the tendency—already discernible by the late 1990s—by telecommunications carriers to evaluate the legal justification of communications interception requests.

The present research also outlines a number of similarities in the British and American telecommunications landscapes, in relation to the impact of market deregulation on the wider dialogue over digital communications interception. In both case studies, the telecommunications industry was segregated between older and younger carriers. The older and more established carriers displayed a sense of reassurance about their historical arrangements with government agencies and about their ability to facilitate state-sponsored wiretapping requests in the digital environment. The younger and less established carriers, however, had ambiguous feelings about assisting law enforcement, and often refused to honor communications-interception requests that were made in the absence of sufficient documentation. Furthermore, the increasing competitiveness that characterized the British and American markets appeared to filter out law enforcement concerns from the day-to-day running of the business of telecommunications. Policing concerns seemed somehow less significant in comparison to the profit motive, which permeated the institutional decisions of telecommunications carriers in a progressively competitive environment. By the late 1990s, that new environment had molded a different generation of private carriers, who were less willing than their state-protected predecessors to prioritize law enforcement and security concerns over their immediate and long-term financial goals.

Equally significant in this research are a number of differences that appear to have marked the communications interception developments in both nations. For example, British carriers were more sensitive than their American counterparts in recognizing that the abuse of communications interception by government authorities could have significant political implications. At the same time, they appeared to be distinctly pessimistic about the technical feasibility of implementing comprehensive wiretapping capabilities in the digital environment. In contrast, having been subjected to CALEA scenarios for almost a decade by 2000, American carriers appeared much more prepared to overcome or bypass some of the more complex technological barriers to state-sponsored communications interception—providing of course that substantial financial resources from public funds were made available to them.

Similarly to the British scene, the telecommunications market environment in the United States was distinctly fragmented and antagonistic. But American carriers were able to have a stronger say in shaping communications interception legislation. That was probably because they had managed to achieve a relative degree of internal unity within the industry, prior to lobbying in government circles. It also appears that the American government was much more willing to take seriously industry and civil-liberty concerns about CALEA. In Britain, the absence of a will for a genuine dialogue on behalf of the government forced the industry to resort to a form of bargaining in negotiations over RIPA. The latter requested added assistance from law enforcement in combating telecommunications fraud, in return for adopting a co-operative stance over the proposed legislation.

7.2 Communications Interception in a Deregulated Market Environment

The scant literature on domestic communications interception indicates that, ever since the dawn of telephony in America and Britain, the lines of demarcation between the telecommunications security and law enforcement communities have traditionally been rather faint. Indeed, on numerous instances, such as during the two World Wars or at the peak of the Cold War, these lines were almost totally obliterated, thus allowing for the establishment of a set of aggressive communications interception policies, many of which were informal. These were ostensibly implemented in order to defend national security. Research conducted in the context of the present book appears to confirm this intimate institutional interface between the government and the telecommunications carriers, and points to the endurance of this phenomenon until fairly recent times. In Britain there is evidence to suggest that law enforcement agencies used what they perceived as a regulatory deficit to conduct unwarranted surveillance of Internet data and content in the 1990s.

But the findings of this research also suggest that the long historical practice of warrantless wiretapping was abruptly terminated at the onset of the telecommunications deregulation of the 1980s and 1990s. The new privatized landscape resulted in a dramatic severance in relations between the telecommunications industry and law enforcement agencies in both the United States and the United Kingdom. Indeed, the degree of deterioration in relations between these two once closely linked actors can hardly be overstated. The decision to end the state-sheltered monopolies of AT&T and Post Office Telecommunications greatly increased the number of actors involved in commercial telecommunications and literally pulled the rug under the feet of law enforcement and security agencies. In the British telecommunications market, for example, which was once made up of a single state-owned monopoly, the industry featured no fewer than 200 public telecommunications operators and several hundred ISPs by 1999. In the United States, where the government's interception agencies were used to dealing with fewer than a dozen AT&T-owned carriers prior to deregulation, there were approximately 37,000 carriers and service providers of all descriptions by 1998, more than half of which did not exist in 1994. The once exclusive club of communications interception (Interview 11 2000, pp. 20–26) disintegrated by virtue of the phenomenal rise in the sheer numbers of its members. These monumental developments remain untraced in the scholarly literature on the subject of information security. The latter has tended to concentrate on official documents, thus missing a number of tacit trends in the day-to-day evolution of communications interception. It was these tacit trends that ultimately propelled CALEA and RIPA onto the foreground.

By the turn of the century, the relationship between the intercepting agencies and the deregulated telecommunications sector remained largely amorphous and unstructured. Moreover, the financial and logistical impact of CALEA and RIPA's imposition on the larger, more established, carriers had a diminishing effect on their relationship with government agencies, especially in the United States. In 2000, an FBI special agent complained that

> as CALEA implementators—whatever that is—we're the last person [the carriers] want to see. When you say CALEA it is like dropping acid in their eyes [. . .]. When it comes to CALEA [. . .] our relationship with the carriers and those who are dealing with all of that has been diminished [and is] adversarial in some cases. CALEA [ha]s really been a detriment to the relations that we previously enjoyed—at least with that small family of carriers—and [. . .] right now I don't see any recovery or normalization of a good and strong working relationship between the carriers and law enforcement. I haven't [. . .] witnessed it. If anything, it's getting more fractionalized—worse [. . .]. We need [the carriers], they don't need us. It's really odd, because [. . .] we're implementing this law and we're creating enemies [. . .] of people that we can't afford to [have as] our enemies. It's really tense [FBI Special Agent, CALEA Implementation Section, Interview 19 2000, pp. 451–453, 809–816, 846–848].

Consequently, by the early 2000s, there was little doubt among industry experts on both sides of the Atlantic that the long-term future of digital communications-interception legislation would be marred by consecutive legal challenges and court battles, which would further-strain relations between industry and law enforcement:

> the RIP Bill seems to proliferate the thinking that 'let's get it into a bill and let the courts decide in the long run'. We suspect—certainly the general feeling on the streets is—that the RIP Bill will get through, but a lot of the issues therein will not be cleared until someone has been taken to court and the highest court in the land has made a case [. . .]. Now, if [the industry] were to take that line, then the system would break down overnight because there is no way that the justice system could go through that process every time one wanted the name of a [. . .] telephone [. . .] subscriber or what have you [Officer, MI5, Interview 12 2000, pp. 108–113, 257–260];

> most of the Internet Service Providers Association, they're ISPs that are big enough, they don't need to bother about the government. They're taking the attitude that 'well, OK guys, you roll it out and we'll test it in court'. The RIPA is a big sledgehammer: 'you will co-operate'. Or will we? It's not very well thought out. Not very well thought out at all [Security Officer, large British ISP, Interview 26 2001, pp. 88–92, 291–292; see also Interview 05 2000, pp. 225–229; Interview 10 2000, pp. 160–164; Interview 11 2000, pp. 163–168; Parr 1999; Dempsey 1995];

> as far as CALEA itself, I think we're just seeing [. . .] an evolving type of law that's going to have challenges after challenges and implementation schedules changed as standards change [Technical Director, national association representing over 500 smaller TSPs, Interview 17 2000, pp. 215–217; see also Interview 13 2000, pp. 216–221; Interview 19 2000, pp. 389–393];

> a year ago, when times were good, everybody leaned towards the view that it was better to not pick a fight with the FBI. Now it's less clear that people have the funds to spend on

development or to purchase this [CALEA] stuff, so there could be a serious conflict over this [Stewart Baker, former NSA General Counsel, quoted in Brown 2001].

On a broader note, it is apparent that, by the late 1990s, deregulation had increased market competition, which in turn had effected an unprecedented restructuring of corporate relations in the British and American telecommunications industries. This new competitive landscape had delivered a serious assault on the ability of centralized governments to police the pattern and content of their citizens' communications. In this sense, and if communications interception is to be seen as part of the overall surveillance capacity of the contemporary nation-state, it would seem that the evolution of capitalist systems is not inherently congruent with the mandates of government-sponsored surveillance. This finding contrasts with the work of surveillance experts such as Ackroyd et al. (1980) or Gandy (1993). Indeed, as this study shows, market deregulation tends to introduce an often-chaotic complexity to rigid state-sheltered systems of communication, and can result in the impairment of surveillance systems and structures that were not designed with deregulation in mind.

7.3 Industry's Elevated Role in Communications Interception

In light of the above observations, it would appear entirely appropriate to suggest that, in a deregulated environment, where governmental intervention becomes structurally limited, the technical surveillance capabilities offered by digitization have not been thwarted as such. In fact, even a cursory study of the contemporary digital telecommunications ecosystem would demonstrate that it directly subverts even the basic tenets of information security. What is more, digital telecommunications systems are far more susceptible to wholesale surveillance—both automated and manual—than analog systems. Therefore, the possibility for highly intrusive mass surveillance is acutely present in the digital telecommunications environment. The difference is that surveillance becomes heavily controlled by corporate actors, to the detriment of state agencies. This tendency has since been amplified by the onset of social media, particularly Facebook (Cadwalladr and Graham-Harrison 2018). This assertion agrees with the overall spirit of capitalist deregulation, which aims to marginalize the role of government in the affairs of civil society (Hills 1986, p. 25). It is also compatible but with the findings of this study in regards to the enhanced role of TSPs in the implementation and oversight of communications interception operations under CALEA and RIPA.

As explained in earlier chapters, CALEA and RIPA interception models propose that, for the first time in the history of communications interception, standardized wiretapping operations would have to be performed in their entirety by the carriers, rather than by law enforcement or security services. Clearly, this shift in the paradigm of state-sponsored communications interception signifies a loss of control by government agencies of the technical process of communications interception. In addition, it would be unrealistic to expect that carriers could assume complete technical responsibility over state-sponsored communications interception without eventually being engaged in a more legalistic role in surveillance operations. Indeed, there was already strong evidence by 1999 to suggest that British and American carriers were beginning, or intended to begin, to query the justification of state-sponsored communications interception requests. Inevitably, then, the carriers' upgraded role in safeguarding their users' information security—which they have always viewed as a revenue-generating feature—was expected to attract considerable pressure from groups or institutions aiming to either increase their communications interception-capabilities or augment the security of their communications. In an important sense, therefore, CALEA and RIPA appeared to recognize the private sector's enhanced control over the information security—or lack thereof—of citizens in the information society.

7.4   Deregulation an Anathema to Communications Interception

In the United Kingdom, the discernible shift in the technical supervision of communications interception from state to industry was partially due to British law enforcement's demonstrated lack of technical knowledge about the specifics of digital communications systems. Indeed, British industry representatives who maintained daily contact with government officials had formed an impression of the latter's technological expertise that varied immensely from popular—and often scholarly (see for instance Manwaring-White 1983)—myths of sophisticated government techno-agents making use of highly advanced interception gadgetry. In reality, at the onset of digitization, British law enforcement officials were considerably under-trained and lacked even basic knowledge about the fundamental features and capabilities of digital networks, including the Internet.

Nevertheless, the aforementioned shift in the control of communications interception was also due to the particular design features—rather than the overall nature—of digital technology itself. These features tended to largely diminish the potential of tampering with the networks at the local loop, or even at the cross-connect box. In other words, there was nothing intrinsic in digital networks that rendered them more resistant to interception than, say, analog networks. Rather it was the absence of law enforcement and security interests from technical decisions about digitization that resulted in the marginalization of their interests from the very design of digital telecommunications systems in the 1980s and early 1990s.

This latter point is potentially more significant than may initially appear: for law enforcement and security agencies, communications interception is not simply another investigative tool. As explained in earlier chapters, there is a long historical tendency within the British and American law enforcement and security communities to regard communications interception as a fundamental pillar of investigative competency. Yet, in spite of that, British and American law enforcement and security agencies appear to have been given a limited and ineffective degree of input in early debates about telecommunications digitization, or even in later debates concerning the standardization of these technologies. When asked about the reasons behind that absence, law enforcement officials replied that it was simply none of their business:

[It all began w]hen they denationalized...when British Telecom lost their monopoly, basically. And...

Q: Didn't law enforcement people sort of forecast that? Or didn't they realize it was going to happen?

A: I mean...no. I mean...they are not part of that discussion [...]. Denationalization and de-monopolization, or whatever you'd like to call it, was created in order to provide market forces and the best commercial aspects for British liberties. Law enforcement doesn't come into that [.... Take, for instance,] the introduction of pre-pay [mobile phone systems]. Law enforcement was very, very much against [them]—had it been given an opportunity. It would have been virtually against by the law enforcement [community]. But, no way, they are not consulted, you see, they are not part of that consultation process [.... So,] whilst law enforcement don't like [denationalization], I think they would agree they're not part of that discussion. It's outside their limit [Officer, MI5, Interview 12 2000, pp. 464–467, 469–472, 486–489, 501–502].

I realize that from a commercial perspective [...] there [wa]s a reason for the break-up of the [AT&T] monopoly. But from an [...] overall quality control and an overall tracking of telephone service it has been very bad and it's getting worse [...]. When they broke up AT&T, the FCC or the Federal Government [...] didn't look at the whole picture. Deregulation on one hand is good for the consumer [...], but on the other hand, for public safety and for the consumer not knowing it was very bad. [T]hat part of it was never looked at. And if you talk to these people they'll say 'oh, yeah, oh gosh we shouldn't have done that'. But nobody investigated this side of it. Nobody asked. The people that were putting the pressure on from the commercial side were putting so much pressure on that the half-paid attorneys and what have you were pushed, pushed, pushed. They didn't tell—they didn't want that to come out because that would influence the court's decision on 'do we need to do this or is it really that good for the market?'. It's been bad for us. At this point in time we just have to fix what we've got [Technical Unit Chief, Emergency Communications Unit, Interview 13 2000, pp. 49–52, 84–85, 97–106].

In the United States even the FBI appeared to recognize the subordination of its perceived security interests to those of a deregulated and commercially-driven national economy within the global free-market environment:

[t]he United States has traded the comfort of uniformity and predictability in its communication system for creative innovation and vigorous competition [...]. The goal of the [CALEA] legislation [...] is not to reverse those industry trends. Indeed, it is national policy to promote competition in the telecommunications industry and to support the development and widespread availability of advanced technologies, features and services [Blair et al. 1995; Boucher and Edwards 1994].

There are strong indications, however, that the relinquishment of national security mandates in favor of capitalist restructuring did not come about smoothly. British academic analyses of telecommunications deregulation contain little evidence of national-security concerns having featured into this primarily economic debate. In the early 1990s, some actors did express concern about the lack of "expressed vision" and of "long-term objectives" in the British government's deregulatory drive (Veljanovski 1991, 20ff). But the absence of national-security concerns from the relevant scholarly literature should not be taken to mean that such concerns were not present. It is far more likely that they were communicated away from the public domain by members of the British intelligence and security communities, or even that scholarly works on the subject simply failed to notice this element of the debate.

In the American context, there are clear indications that at least some segments of the nation's security establishment wasted little time in publicly and strongly opposing the intentions of the Reagan administration to break up AT&T in the interests of competition (Judge 1985). As early as February 1981, only days after moving into the Pentagon, United States Secretary of Defense Caspar Weinberger drafted an urgent memorandum to the United States Attorney-General William Smith, in which he expressed in his Department's vehement objections to the antitrust lawsuit against the AT&T:

> [t]he purpose of this letter is to express the deep concern which the Department of Defense feels over the reports of the proposed settlement of the government's old antitrust suit against the American Telephone and Telegraph Company [. . .]. Our concern is based upon the fact that a great deal of the current capability for communications command and control of our strategic weapons depends upon the continued existence of the only communications network in the United States capable of providing the services required [. . .]. The Department of Defense recommends very strongly that the Department of Justice not require or accept any divestiture that would have the effect of interfering with, or disrupting, any part of the existing communications facilities or network of the American Telephone and Telegraph Company that are essential to defense command and control [quoted in Carter 1989, p. 224; see also Tunstall 1986, p. 104].

Weinberger reiterated his Department's concerns a month later during testimony before the United States Senate Armed Services Committee, where he stated that

> [t]he American Telephone and Telegraph network is the most important communication net we have to service our strategic systems in this country. Because of the discussions I have had concerning the effect of the Department of Justice suit that would break up part of that network, I have written to the Attorney General and urged very strongly that the suit be dismissed, recognizing all of the problems that might cause and because of the fact it seems to me essential that we keep together this one communications network we now have, and have to rely on (quoted in Carter 1989, p. 224).

The Defense Department's strong objections against the AT&T divestiture was also communicated by the then United States Deputy Secretary of Defense, Frank C. Carlucci, in a letter addressed to the United States Assistant Attorney-General William Baxter, who was also head of the Department of Justice's Antitrust Division. In it, Carlucci urged Baxter to consider the "severe problems [that will] confront the Department of Defense if this network is broken up" (quoted in Carter 1989, p. 225). In 1981, Carlucci helped compile an additional report by the United States Defense Communications Agency, which effectively became declassified upon its inclusion by AT&T's legal team in the antitrust trial. The report stated that the Department of Defense was in a position to unequivocally state that divestiture [. . .] would cause substantial harm to national defense and security and emergency preparedness telecommunications [because it] would substantially reduce or eliminate entirely the incentives [. . .] to engage in that prior joint [—among different service providers as well as between providers and government—] planning and preparation [that is necessary] to conduct centralized network management. The Defense Department totally disagrees that divestiture would have no adverse effect on the nation's ability to rely upon the nationwide telecommunications network. Instead, we believe that it would have *a serious short-term effect and a lethal long-term effect*, since effective network planning would eventually become virtually non-existent (quoted in Bolling 1983, pp. 53–54, emphasis added).

The Department of Defense's lobbying convinced President Reagan to create the presidential-level National Security Telecommunications Advisory Committee, which initially included the heads of 27 telecommunications industry members and today continues to advise the Department of Homeland Security on issues deemed critical for security and emergency preparedness. Yet, in regards to deregulation, the Pentagon's lobbying efforts left Congress and the FCC essentially unmoved (Carter 1989, p. 224). Eventually, the Reagan administration decided to implement the break-up of AT&T against the advice of the nation's defense and security communities. Observers noted at the time that deregulating the American telecommunications sector was of immense symbolic significance for the Reagan administration's broader liberalization program (Hills 1986, 24ff). It also appears that the plan's successful completion was considered more politically expedient than the protection of equally important—yet less prominent and less voter-worthy—national-security mandates. That prioritization would appear to lead us to the broader conclusion that, when it comes to the surveillance and policing of telecommunications, strategic coherence between the different sectors of the state should not be assumed. In fact, some writers point toward policy inconsistencies within the Department of Defense itself. Many of its senior officials opposed the deregulation of telecommunications, which they saw as adversary to information-security goals, while at the same time favoring the global liberalization of telecommunications. The latter was seen to be furthering the American desire to control transnational fiber optic networks and the communications-hardware export industry (Hills 1986, p. 195; Tunstall 1986, p. 55).

Admittedly, government policies relating to telecommunications and security tend to be shaped through the clash of mutually competing interests and even "turf battles" (Pfaltzgraff 1984, p. 297). These policy struggles tend to give shape to legislative aspirations without necessarily considering their impact on the state's long-term strategic mission (Pfaltzgraff 1984, p. 292; Krasner 1978; Jepperson et al. 1996, 50ff). In this particular instance, what some policy-makers perceived as economic security came in direct conflict with national-security directives. Thus, market deregulation in the field of telecommunications resulted in serious destabilizing repercussions that directly affected what the law enforcement and security communities saw as one of their most valued investigative weapons. Ultimately, in this case, the state's ability to access the patterns and content of citizens' communication was seriously thwarted, not due to some lobbying onslaught by civil-liberties activists, nor because of large-scale sabotage by politically-minded computer hackers, but by the internal inconsistencies of government policy.

7.5   Blackboxing the Communications-Interception Debate

There are few—if any—issues in the general debate on domestic communications interception that justify a genuine need for concealment. It is true that a very small number of individuals are actively involved in the practice of communications interception. However, developing basic knowledge of telecommunications systems and consulting publicly available technical manuals are usually sufficient to familiarize oneself with the inner workings of communications-interception models and techniques. Even highly classified interception software and hardware, such as FBI's DragonWare, were in fact by-products of commercially available monitoring tools, all of which remain available to consumers. In the United Kingdom, law enforcement officials themselves routinely recognized in the 1990s that criminals wishing to evade law enforcement communications interception schemes were "already aware of [law enforcement's] capabilities" (Stoddart 1999). Yet, despite that remarkable openness—or perhaps because of it—secrecy has been an indispensable component of both the British and American communications-interception debates. The present research appears to confirm the findings of the wider literature on the subject, which tend to view secrecy as an inherent feature of the culture of state-sponsored communications interception.

There is little doubt that national-security issues, such as state-sponsored communications interception, tend to be insulated from popular control and knowledge, even in representative democracies (Russett 1990, p. 146; Interview 05 2000, pp. 399–406; Interview 08 2000, pp. 453–454; Interview 11 2000, pp. 446–449). Nevertheless, no evidence has emerged in the course of this research that would lead to an interpretation of secrecy as the primary—or even as a systematically planned—barrier to accountability and transparency on matters relating to state-sponsored communications interception. To begin with, secrecy has not been unilaterally imposed upon the communications interception debate by the British and American state actors. In the course of this research, corporate secrecy and concealment proved much more rigid and impenetrable than governmental attitudes. An indicative example is that, on both sides of the Atlantic, the ratio of interview requests to affirmative replies was much higher in the case of government officials than in the case of corporate representatives. Furthermore, it has already been noted that, even prior to the emergence of IOCA—RIPA's short-lived predecessor—in the United Kingdom, TSPs were instrumental in concealing the digital communications interception debate beneath a veil of secrecy. Eventually, this veil was respected even by a number of civil-liberties groups that participated in negotiations over RIPA (Bowden 2000, pp. 296–300). Thus, by no means were restrictions imposed on the CALEA and RIPA negotiations unilaterally by government officials. Rather it was an all-embracing cultural attitude that was rooted in market competition between carriers, as much as in enduring bureaucratic values that lie at the heart of the modern nation-state.

More importantly, the very fact that the present study was accomplished—albeit with some historical distance—testifies to the considerable amount of communications interception-related information that has been made public by government agencies on both sides of the Atlantic. In fact, the degree of government openness on CALEA and RIPA has been unprecedented both in the United States and the United Kingdom—where until relatively recently the government often denied even the existence of its co-operation with industry on matters relating to communications interception (Interview 12 2000, pp. 451–455). Indeed, the communications interception agenda of American and British law enforcement and security agencies was widely advertised in the public domain in the late 1990s, either deliberately—such as by the FBI's public-relations campaign to promote CALEA via its, now defunct, www.askcalea.com website—or through decisive document leaks in the press. An example of the latter was the leak of *Looking to the Future*, a restricted document that was jointly drafted by representatives of several British intercepting agencies and addressed to the Home Office. Notably, the document recognized that RIPA's communications-interception requirements subverted human-rights legislation as sanctioned by Data Protection laws and the European Union (Gaspar 2000, Sect. 7.1.1), and yet refused to redraft RIPA. Furthermore, the document proposed further law enforcement and security-friendly communications interception mandates, including a suggestion that all digital and analog data "generated or routed through a TSP's network" by every user in the country should be retained for a period of 7 years (Gaspar 2000, Sect. 6.1ff). It also demanded that the British government

> be prepared to defend our position, accepting that once communications data has been used to satisfy the business needs of communication service providers, retention is still vitally important to the Agencies and the Criminal Justice System [Gaspar 2000, Sect. 7.1.3].

The document was leaked to the *Observer* newspaper (Ahmed 2000), where it made front-page news and generated numerous editorials in the British press. The fact that, despite the revelations of *Looking to the Future*, RIPA was proposed and eventually enacted by the British government without causing widespread political turmoil among the British electorate is significant in its implications. It should lead researchers to question the simplistic view that secrecy is what stands between the machinations of conspiratorial government bureaucrats and an ignorant yet volatile electorate, which is prepared to defend civil rights as soon as they are threatened by the policies of non-elected administrators.

Chapter 8
Epilogue: Surveillance in the Information Society

Undoubtedly, CALEA and RIPA represented organized attempts by government agencies to gain access to the design of public telecommunications networks, which had previously developed in the absence of collateral principles, such as national security or law enforcement interests. The shared ambition of CALEA and RIPA was to elevate the status of such principles to guiding parameters in the design of digital telecommunications networks—an attempt that has been described by observers as "forced integration of specific enhanced surveillance features into the nation's telecommunications infrastructure" (Dempsey 1998).

Early indications suggested that the United Kingdom's RIPA interception model favored a vision of a permanent interception presence built into the digital telecommunications network. Yet, even in the United States, where advanced models of ISDN interception appeared to be based on detachable wiretapping technology, industry representatives were detecting an element of technical permanence in CALEA blueprints, which they saw as introducing "a surveillance functionality into the [...] switching fabric and into the logic" (Interview 16 2000, pp. 229–230) of telecommunications networks. It is important to stress here that, prior to digitization, law enforcement and security agencies never requested to have an input in the design of telecommunications systems in the American and British contexts. In 1994, two American members of Congress noted that

> until recently, the question of system design was never an issue for authorized surveillance, since intrinsic elements of wirelined networks presented access points where law enforcement, with minimum assistance from telephone companies, could isolate the communications associated with a particular surveillance target and effectuate an intercept (Boucher and Edwards 1994)

Yet all that changed due to digitization, and the problem was compounded by the deregulation of telecommunications. That led to proposals to redesign all digital switches, which was a minimum requirement in both CALEA and RIPA. The American experience showed that carriers were able to delay the redesign of switching equipment on financial and administrative grounds. However, by 2001 it

© Springer Nature Switzerland AG 2020
J. Fitsanakis, *Redesigning Wiretapping*, History of Information Security,
https://doi.org/10.1007/978-3-030-39919-1_8

appeared certain that the next generation of switches would come with standard ports specifically designed for communications interception (Interview 11 2000, pp. 129–132; Interview 18 2000; Blair et al. 1995). In the ensuing years, the realization of that redesign signified the first documented instance where American and British law enforcement and security agencies were permitted to have an input into the design of public telecommunications systems equipment, and even features. Thus, telecommunications networks were—gradually and in stages—transformed into vehicles for the practical expression of law enforcement and security management goals. In an important sense, therefore, the long-term outcome of CALEA and RIPA has been to subvert the information security architecture of the telecommunications system in order to facilitate state-sponsored communications-interception mandates.

8.1   A Political Economy of Information Security

This unprecedented development is ongoing and has yet to be fully examined in the scholarly literature. We are obligated, however, to assess its full meaning for information security as a discipline. In fact, if we were to interpret the history of British and American telecommunications regulation based solely on the available scholarly literature, we would have to infer that information security—and, to a lesser extent, national security—concerns have been almost completely absent from the decision-making process. The present analysis, however, demonstrates that such an inference would be highly inaccurate. Information security and national-security directives were certainly not singular influences on the formation of information security policy; yet their role was anything but trivial and often helped shape the state's administrative perspective on telegraphy and early telephony.

These influences were not abandoned following the paradigm shift from electric to digital modes of communication. They are very much alive today, in what is often called the era of the information society. More than two decades ago, as the world witnessed the dawn of digitization, national-security concerns prompted the emergence of CALEA and RIPA, two major pieces of legislation that changed the very design of information security in the context of digital telecommunications. The technical changes that were triggered by the legislation inevitably lead us to a sociotechnical analysis of information security in the digital domain. This is the view that the information security features of telecommunications networks inevitably embody the sociopolitical visions—including predispositions and biases—of their designers. In light of the findings of the present research, therefore, future attempts to assess the information security aspects of the information society must take into account their sociotechnical particularities. In the words of Professor Robin Mansell, of the London School of Economics, we need to evaluate the information security of telecommunications from a sociopolitical point of view. Along with others (Clark et al. 1989; Molina 1990), Mansell argues that a political economy of network design must be implemented [with the aim] of exposing aspects of telecommunications network evolution that bias its development and become part of a complex system of institutionalized power relations [...]. The political economy of the telecommunication network will offer an analysis, which is complementary to related approaches to the relations of power and control that are embodied within the electronic communications environment. It should also focus on choices regarding the way in which network evolution influences calling patterns, privacy of customer information, and the centralization or decentralization of control over information conveyed, or generated by public network transactions (Mansell 1993, p. 35).

A significant benefit of analyzing information security under the prism of sociotechnical discourse is that it facilitates the systematic consideration of non-technological factors in the evaluation of information security. This, in turn, allows for more accurate forecasts of the trajectory of communications-interception trends in an increasingly digitized and deregulated telecommunications environment. In other words, any assessment of ongoing trends in information security must recognize that strictly technical developments in telecommunications will not in themselves be decisive in determining the future of communications interception.

Digitization is an illuminating case in point: regardless of whether digital telephony systems in the United States and the United Kingdom will move toward a wireline (fiber optic) or a wireless (satellite-based) future, digitization will continue to subvert state-run communications-interception models. In this sense, even in their present stage, neither CALEA nor RIPA are future-proof or able to guarantee the exercise of state-sponsored communications interception in the future. This was noted nearly 20 years ago by several actors involved in communications-interception policy:

> with ISDN coming up, with fiber to the home, I mean, what are you going to do then? You can't go inside a person's home and run some beads out the back door, you know, into a recording device [Director, National Services Planning, large TSP association, Interview 37 2000, pp. 32–34];

> with fiber-to-the-home [. . .] there's no switch point [. . .], there's no place to put [a wiretap]. Are you going to have [. . .] the government agency knock on your front door and say 'I want to tap into your fiber system'? It [. . .] isn't going to happen. So, yes, I think there are more questions that are going to be asked and there are very few solutions that are going to come forward [Technical Director, national association representing over 500 smaller TSPs, Interview 17 2000, pp. 208–215].

Back then, fiber-to-the-home technology was but one potential element in the development of digitization that led many observers to question the future technical feasibility of wiretapping practices. An interviewee specializing in American telecommunications legislation confided:

> I think the interesting thing about legislation and regulation [. . .] in that sort of technology world is that technology is constantly outpacing regulations and legislation. And you can see with CALEA that [. . .] they never foresaw the [. . .] problems they're having with [information technology]. By the time the FCC gets a rule out on anything technology has already far surpassed it and it's already dated. I have this feeling that by the time all of this gets sorted out in the courts they're going to have a whole new set of questions to answer [. . .] as we have all this convergence going on in the industry (Government Relations Director of a large

American telecommunications association, Interview 18 2000, pp. 75–181; see also Interview 19 2000, pp. 531–532; Freeh 1997).

British interviewees were equally pessimistic about RIPA's prospects of endurance:

I think maybe in 10 years' time we'll need another RIP Bill [author's note: we did], which will kind of merge it, so that the whole thing becomes a bit more technology-neutral then (Law Enforcement Liaison, large British ISP association, Interview 29 2001, pp. 22–24);

there is a need to change all this, but then, the question arises 'well it's all going to change again [. . .] within the next 3 or 4 years' [laughs]; how do you write legislation which has got to be future-proof, but is not just a blank check for the Secretary of State [. . .] to do what he feels like at the time? (Head, Regulation Affairs Department, large British TSP, Interview 32 2001, pp. 290–294);

there's nothing there behind [RIPA] on which we can put a yardstick and say 'yes, that will actually sit on and stand the test of time' (Officer, MI5, Interview 12 2000, pp. 297–298; see also Interview 04 2000, pp. 265–272).

It is certainly true that, even as these interviews were conducted in 1999 and 2000, parts of CALEA and RIPA appeared to have already been surpassed by technical developments, such as VoIP. However, as this research has shown, communications interception in the ISDN, fiber optic and even packetized data environment has always been technically feasible—for instance through bridges installed in local area networks, through Internet protocol detection and identification systems, or through traffic mediation and filtering. Another comprehensive communications-interception scheme, which has never been publicly suggested, though it is entirely feasible in the technical sense, could involve the reconfiguration of the national telecommunications network's routing structure, which would make it mandatory for all information transmitted through the network to pass through any one of a number of designated switching stations equipped with mediation and filtering devices. As technology continues to develop, this could become a future trajectory in twenty-first-century communications-interception models.

Nevertheless, the fact that digitally enhanced models of communications interception are technically feasible does not mean that they will materialize. As shown in this study, financial implications are likely to be of paramount importance for carriers when it comes to deciding on the particular technical features of digital communications interception. In the words of one industry observer, "anything is technically feasible. [The question i]s at what price is it feasible and who's going to pay for it?" (Interview 07 2000, pp. 160–161; see also Interview 01 2000, 349ff). Ultimately, whether governmental communications interception ambitions will be realized in the digital age will largely depend on the potential impact of these ambitions on the financial viability of commercial carriers.

Additionally, this research strongly suggests that cultural forces have played— and will continue to play—a pivotal role in determining the operational limits of communications interception in the digital ecosystem. As explained in previous chapters, the deregulation of the telecommunications market resulted in an unprecedented cultural shift in corporate culture in both Britain and the United States. Thus, although the new, private TSPs recognized the social significance of telecommunications policing in the 1990s, they did not appear to share their predecessors' favorable attitude toward the operational mandates of law enforcement and security agencies. As this study has demonstrated, the traditionally close relationship between British and American law enforcement agencies and TSPs has evolved over nearly two centuries. That evolution has taken place largely in the shadows of legal accountability and is the historical product of complex political and institutional factors. During the negotiation phase of CALEA and RIPA, it was apparent that any efforts to restore that relationship in the deregulated environment would indeed benefit by operational proximity prompted by strong legal norms. However, that restoration was likely to take place mostly outside the realm of legal directives. Ultimately, without the restoration of that relationship, the realization of the full extent of CALEA and RIPA's ambitions for effective communications interception will prove unattainable in the deregulated telecommunications environment.

8.2   The Effects of the 9/11 Attacks

The above analysis has significant implications for the future of state-sponsored communications interception as a concept and as a practice. It means that the technical parameters of communications interception in the digital environment will not determine the nature or extent of its use. In the course of this research, it became clear that TSPs in both Britain and America were certain that the technical reconfiguration of their network, as required by CALEA and RIPA, would result in an inevitable increase in the volume of intercepted traffic (Interview 18 2000, pp. 57–46; Interview 20 2001, pp. 232–237; Anonymous 1997a, 1999b; Dempsey 1995; O'Doud, 1999; Leahy 1999). In the United States, service providers privately described the potential volume of intercept requests as "huge", "unreasonable" and "astronomical" (Auble and Bender 2000). In 1997, Thomas Wheeler, President of the Cellular Telecommunications and Internet Association, said about CALEA: "all we wanted was to move from a propeller-driven airplane to a jet airplane, but the FBI wants to build the Apollo program" (quoted in Vesely 1997). The technological suggestions that underpinned CALEA and RIPA appeared to vindicate such expectations: the permanent hardwiring of interception capabilities into digital switching technology was bound to be a challenging task because of the sheer number of switches that would have to be modified and replaced. However, once this task was completed, the permanent presence of communications interception techniques in the network, was likely to reduce the financial cost of wiretapping operations: the need to dispatch government agents out into the field to install wiretaps would be eliminated; the interception and collection functions of communications-interception practices would be simplified; and the processing of intercepted data would be automated, possibly down to the transcription of oral communications.

But the technical ease alone could never in itself be sufficient to facilitate increased communications interception by state agencies. The only thing that could do that would be a strengthening of the institutional interface between law enforcement and the new, deregulated telecommunications carriers—something that both industry and government representatives indicated in the context of this research. There is strong evidence to suggest that the dramatic events of 11 September 2001, served to significantly strengthen the institutional interface—both formal and informal—between government agencies and the telecommunications sector in the United States and Britain. Specifically, the radical changes in the regulation of information security that the United States and Britain experienced following the attacks of 9/11 were instrumental in eliminating resistance by telecommunications carriers to communications-interception requests.

In Britain, the security chill felt by the 9/11 attacks was compounded by the London bombings of 7 July 2005, when a series of coordinated suicide attacks on the city's public transport system killed 52 people and injured hundreds more. These events prompted a new information-security paradigm in the country and propelled RIPA into becoming a central investigatory tool for law enforcement and security services. In the year following the London bombings, more than 1000 RIPA applications were made every day by a variety of government agencies in the country (Rayner and Alleyne 2008). In addition to a higher frequency of communications-interception requests, RIPA also facilitated a striking expansion in the number of government agencies that were entitled to request access to communications data. In 2000, when RIPA was introduced in Parliament, approximately nine government agencies were entitled to make use of it. By 2009, that number had risen to 800 and included hundreds of local councils (Dodd 2009; Porter 2009). Consequently, in 2007 alone, RIPA was used more than 10,000 times by local councils around Britain. Many of these requests were reportedly issued to assist investigations into relatively minor offences, such as underage smoking and illegal waste-dumping (Savvas 2008).

In America, CALEA's scope has been greatly expanded to include all VoIP and broadband Internet traffic. Attorney Konrad Trope has noted that, from 2004 to 2007, "there was a 62 percent growth in the number of wiretaps performed under CALEA, and more than 3000 percent growth in interception of Internet data such as email" (Trope 2014, p. 3). Much of that reflects Americans' increased use of the online environment since CALEA's initial implementation. But it also reflects the political pressures caused by the 9/11 attacks. Criminal justice scholar William Bloss has demonstrated that the way in which the sense of insecurity caused by the 9/11 attacks gave rise to the concept of "preventive law enforcement'; a central function of this preventive approach was facilitated by CALEA, as the legislation played a central part in "giving the police broader surveillance powers', says Bloss (2007, p. 209). Soon after 9/11, CALEA was paired with additional pieces of legislation, including the Uniting and Strengthening America by Providing Appropriate Tools Required to Intercept and Obstruct Terrorism (PATRIOT) Act of 2001. These new regulations compelled TSPs to offset the police and security services' "lack of manpower [sic] and technical expertise to keep pace with global terrorists and criminals" (Bloss 2007; see also O'Harrow 2005, 23ff). The ensuing President's Surveillance Program (PSP), which was authorized by President George W. Bush shortly after the 9/11 attacks, formed the legal basis of the NSA's STELLARWIND program, a large-scale effort to data-mine the private communications of American citizens, with the help of TSPs. When the program was divulged to the media through a series of disclosures by whistleblowers such as Thomas Tamm, Mark Klein, Russell Tice and William Binney, the government brought charges against them while at the same time taking legal steps to protect TSPs from privacy-related lawsuits (Aid 2009, pp. 288–293). The fallout from STELLARWIND, which critics condemned as unconstitutional (Hayden et al. 2014), caused waves of protests against the government's surveillance practices. They prompted many to recall the illegal surveillance programs that the American Intelligence Community was found to have authorized for many decades in the lead-up to the Watergate scandal, and which contributed to what one knowledgeable observer described as the "historic breakdown of [American society's] Cold War consensus" (Olmsted 1996, p. 4).

As can be expected, the practical extent of communications interception in the post-9/11 environment will continue to depend on legal responses to technical innovation. The design and use by government of more aggressive communications-interception techniques has been met with legal challenges by carriers and civil liberties watchdog groups—especially as the uses of these techniques appear to urge a renegotiation of previously enacted information-security and human-rights legislation (Landau 2010, p. 2). Henceforth, the outcome of future legal challenges will greatly depend on the political context of national and international relations. Even before 9/11, it was apparent that events that occurred in the context the Cold War, or domestic paramilitary activity—such as the Vietnam War and the Oklahoma City bombing in the United States, or the Northern Ireland dispute in the United Kingdom—triggered more aggressive uses of state-sponsored communications interception (Anonymous 1996; Dillon 1999). What is more, as shown in this study, American law enforcement agencies have shown a propensity to employ political events like 9/11 to justify their communications-interception mandates (Strum 1999). In this sense, one would have to conclude that the turn of events linked to ongoing issues such as the political fragmentation and instability caused by Brexit, the revival of a "cold war" between the United States and Russia, as well as with China, or the seemingly never-ending "war on terrorism", are inextricably linked with the future of communications-interception regimes in both countries. As was remarked in the introduction to this book, the reliance of law enforcement and intelligence agencies on communications interception increases in periods of existing or perceived crises. During such periods, the information-security features of telecommunications networks tend to be systematically subverted by both state agencies and service providers, often with little regard to technical constraints, legal parameters or ethical doctrines.

8.3 The Pluralism of Information-Security Policy-Making

The technical vision of CALEA and RIPA serves to discredit the popular myth that the decentralized architecture of digital communications networks, on which the concept of the information society is based, is inherently uncontrollable by governmental entities (King 1984; Masuda 1981; Horner 1996; Negroponte 1995). Nevertheless, it should not be assumed that the conception, emergence and implementation of CALEA and RIPA were in any sense straightforward, delineated processes. Importantly, both pieces of legislation were enacted *following* the standardization and implementation of consumer-oriented telecommunications services, which attests to the exclusion of law enforcement and security agencies from these processes. Instead, both in the United States and the United Kingdom, the telecommunications digitization debate was observed and steered by financial-oriented state agencies, whose priority was to promote the competitive nature of the market, rather than to enhance, or even maintain, the policing capabilities of the state apparatus. It was only at a much later stage that law enforcement and security officials were able to exercise the degree of influence that was necessary to alter the course of the state's telecommunications policy, and thus steer it to a surveillance-oriented direction. In both cases, therefore, the state appears, not as a unified and monolithic decision-making mechanism, but rather as a field of struggle between pluralistic forces who compete for access to state decision-making and planning.

The above assertion should not lead to a view of the national security community as an actor that is isolated from state structures in representative democracies. In an important sense, the aggressive promotion by state officials of CALEA and RIPA, as well as their eventual implementation, demonstrates the ideological and functional proximity between executive decision-making on the political level and the law enforcement and security establishment in representative democracies. Particularly in the context of the RIPA debate in Britain, telecommunications industry and civil liberties observers appear to have been highly ineffective in influencing political decision networks. The latter appear to have been almost exclusively occupied with national-security priorities and concerns:

> whether you're talking about the interception of communications or acquisition of communications data or decryption powers or oversight or whether you're talking about the boundaries between those or whether you're talking within those fields themselves—wherever there's been a benefit of doubt or some leap that's got to be made to bridge the old regime and the new regime, the benefit of doubt has gone to law enforcement [...] to a quite astonishing extent, throughout the [RIP] Bill [Bowden 2000, pp. 67–72, 186; see also Hall et al. 1999].

There is little doubt that, both in the United Kingdom and the United States, law enforcement and security institutions used their status as members of the state apparatus to engage in what is often called "direct lobbying"—that is, lobbying that is directed toward members of the executive cabinet and departmental officialdom (Shaiko 1998, p. 261). Despite their systematic lobbying efforts, however, the vocal and traditionally powerful American law enforcement and security communities were unable to have CALEA technically implemented until fairly recently, nearly 20 years following its enactment. Resistance to the legislation by the American telecommunications industry and its political allies has been impressive and unprecedented, and has seriously threatened law enforcement's wiretapping capabilities with extinction. There is no evidence that the British law enforcement community's drive for a wiretap-friendly telecommunications environment has been more expeditious: it was only after 2009, nearly a decade after RIPA's enactment by Parliament, that adequate technical communications-interception standards began to emerge from ongoing negotiations between industry and government (Policy Engagement Network 2009, p. 17).

As of 2018, RIPA had gone through five overhauls, which took place in 2003, 2005, 2006, 2010, and 2015. Such administrative impediments attest to the existence of an intricate labyrinth—a complex network of actors—that stand between the state's national-security directives and their eventual application by a highly deregulated, competitive and evolving industry. What used to be a relatively closed-off sociotechnical constituency has now burst open and is continuously subjected to the eventful arrival of new entrants, such as Facebook, Wire, Skype, WhatsApp, and many others. These new arrivals are characterized by different administrative priorities and cultural norms. The degree of conflict—or, perhaps, the degree to which conflict is visible to external observers—has thus increased, and in doing so has further destabilized the formation of communications-interception policy.

8.4 Sociotechnical Trends in Communications Interception

Nevertheless, as explained earlier, a host of sociotechnical factors will ultimately determine the future trajectory of communications-interception regimes. The financial implications of communications interception are of primary importance for industry actors, as the latter continue to operate in a highly deregulated market environment. Competitive-oriented carriers have been unwilling to cover the cost of technical modifications, which means that law enforcement and security agencies will have to continue to lobby political decision centers for the allocation of extensive financial resources. That is true especially today, as the political expediency generated by the attacks of 9/11 has largely dwindled. Additionally, it could take years—possibly even decades—for cultural relations between the new, privately-owned TSPs and security-oriented government agencies to strengthen enough to accommodate the full extent of CALEA and RIPA's operational mandates.

Essentially, the answer to the question of whether state-sponsored communications interception will continue to be enhanced in the digital environment is twofold. Enhanced communications interception could be defined as the construction and application of technical systems that are able to intercept, collect, store and analyze information in a more intrusive, detailed and timely fashion. If we accept that definition, then the state's ability to intercept should be expected to continue to increase by virtue of digitization. On the other hand, enhanced communications interception could be defined as the ability of organized government to secure: (a) a continual and effective technical interface with intelligent communications networks, which are owned by private corporations and whose technical features constantly evolve; and (b) the establishment and maintenance of consensus among disparate corporate actors in accepting law enforcement and national security mandates underpinning communications interception policy. If we opt for that definition, then the state's ability to intercept in the digital, deregulated telecommunications environment will undoubtedly continue to face serious limitations.

This should ultimately lead us to the broader observation that the social shaping of technology does not necessarily imply a smooth and horizontal dynamics between existing or emerging social values and the process of technological design. As new technological paradigms emerge, they are drawn into the arena of political negotiation and become subject to a host of contradictory requirements by disparate actors. The nature of these requirements is formed by the actors' perception of the challenges that new technologies may pose for their particular interests, commitments and goals. The latter are often contradictory to those of other actors. Inevitably, the nature and extent of digital communications interception should be expected to evolve in relation to these contradictory forces.

Ultimately, the legislative framework of communications interception is open to interpretation over time, and its implementation remains highly dependent on the political challenges posed by political events—with 9/11 being just one example among many. An FBI special agent characteristically remarked that "CALEA [. . .] is sort of like the Bible: every time you read it you can see something new in it—a new interpretation" (Interview 19 2000, pp. 750–752). Numerous observers have noted that both CALEA and RIPA are vague and fail to define important expressions and terms, such as "public telecommunications network", "carrier" and "reasonable intercept capability" (Berman and Dempsey 1996). Nevertheless, according to law enforcement officials, this legislative vagueness is not a conceptual failure, but rather a strategic omission, which is intended to accommodate the possible modification replacement of the legislation, as various technical, social and political conditions change:

> there should be no restrictions placed on the scope and type of communication services which are subject to interception legislation. If exemptions are made, then the criminal will exploit them. By leaving the legislation wide, future developments and innovation will fall under the legislation [Todd 1999].

This can be taken to imply that the meaning of expressions such as "reasonable intercept capability" is to be interpreted according to the immediacy of perceived threats to the social stability, economic wellbeing and overall security of the nation state. During the Cold War, for instance, the extent of legal and illegal state-sponsored wiretapping increased as concerns over the activities of real or imagined internal adversaries overtook governmental administrations in the United States and—to a lesser extent—the United Kingdom. In the lead-up to 9/11, there was very little consideration of the parameter of socio-political stability in the debate over digital wiretapping. Civil liberties lobbyists, however, were clear that it needed to be considered:

> even if we trust the current government that doesn't necessarily mean that we will trust all future governments. I mean, MI5 in the past monitored Jack Straw [and other members of the British] Parliament. Phones were listened to on a regular basis. So, the government should be very careful and think twice before passing [this] piece of legislation [...]. Today we have the Labour government [..]; tomorrow [...] Tony Blair might be target of this sort of surveillance [Akdeniz 2000, pp. 227–234];

> each successive generation says 'ah, yes, well that was in the past, that was 20 years ago, of course things are not done in that way right now'. But I think one really has to view this approach with some skepticism and question the assurances that one gets today. [One ought] to worry about how these powers could develop in the future or how they could be abused by a more extreme government [Bowden 2000, pp. 273–275, 384–385].

In the past two decades, the events of 9/11, Brexit, as well as the rise of extremism and populism in Europe and America, have demonstrated that the pillars of socio-political stability in the West are not as sturdy as they are often perceived to be. As the CALEA and RIPA legislations continue to evolve, the political context that surrounds their implementation may well prove paramount.

8.5 Questions for Future Research

In addition to the above conclusions, the present book demonstrates that governmental and corporate secrecy over communications interception is not a valid justification for the widespread absence of vital research on the field. Persistence and networking on behalf of researchers can penetrate the mistrust of external observers—including academics—that government and industry officials may initially display. Undoubtedly, the issue is controversial; yet that should not be seen as a negative element from the viewpoint of research. The lack of substantial agreement in the debate often generates frustration among the principal actors, which in turn makes members of every side in the debate eager to communicate their own story to the wider public.

Neither should the constantly evolving status of digital communications-interception legislation frighten researchers off the field. On the contrary, it should attract them to it. The incompleteness of the debate offers a unique opportunity for interested observers to glimpse into the chaotic construction site that is the process of sociotechnical negotiation between different policy actors. Usually, once the debate has been concluded and the negotiation process has subsided, the polished façades of industry's public-relations exercises and government's political rhetoric will deflect academic interest and curiosity. Researchers then have to engage in a sort of archaeological excavation of the debate, which is usually much more laborious than having had the luxury of witnessing the process of policy formation in the first place.

One such archaeological excavation is required to address the gaps in our knowledge of the history of government telecommunications regulation, as well as of the history of wiretapping in the United Kingdom and the United States. In the case of the former, the impact of national-security concerns on telecommunications policy is totally absent from the scholarly literature; the latter has tended to reduce government regulation into a purely economic affair. Additionally, although the use and abuse of state-sponsored communications interception has been aptly documented, the scant literature on the subject tends to analyze the phenomenon as a linear and conflict-free aspect of national security. Surprisingly little attention has been given to issues of power, adversity and negotiation between government and industry actors involved in the conception, implementation and application of information security in telecommunications. The exposure of conflict in contemporary debates over digital communications interception should not necessarily be viewed as a recent phenomenon. On the contrary, it should be assumed that many historical accounts of the practice have failed to expose and draw attention to the inner complexity of the debate, which has probably characterized wiretapping ever since the beginnings of electric telegraphy.

The future of digital communications interception is also worthy of researchers' attention: if the fragmentation of the telecommunications sector continues under deregulation, then the rising numbers of participants in the communications interception debate should be expected to impend tendencies by government and industry to monopolize access to relevant information. Thus, as the debate over state-sponsored communications interception opens up to public scrutiny, the role of research should be to deconstruct issues of trust between civil society and government, as well as evaluate the impact of that trust—or lack thereof—on emerging models of information security. On the other hand, if the fragmentation of the telecommunications sector is overtaken by oligopolistic tendencies—mergers, etc.—in the relative absence of state interference, then the potential of the debate being black-boxed will increase. Faced with such prospects, scholarly research will have to become more inquisitive and demand access to elements of the communications-interception debate that are purposely kept beyond the reach of public scrutiny.

Another potentially unpredictable factor in the debate is the increasing globalization of the telecommunications market. Indeed, the fragmentation of the telecommunications sector has only been augmented by the increasing globalization of telecommunications and the emergence of colossal media conglomerates (Iosifidis 2014, p. 428; Noam 2014, 503ff). By 2016, the United States government had taken the unprecedented step of banning two Chinese telecommunications manufacturing companies, Huawei and ZTE, from bidding for government contracts in the United States. The reason given was that the two firms—both of which feature on the list of the world's five largest telecommunications hardware manufacturers—"might undertake industrial and strategic espionage for China" (Jiang et al. 2016, p. 34). Both Huawei and ZTE are included in the eight Chinese companies with direct investments in Britain's telecommunications infrastructure. In 2016, the British government forced Huawei to fund the Cyber Security Evaluation Center, from where British cyber-security experts monitor Huawei's products for espionage and security-risk potential (Jiang et al. 2016, p. 37). But, as the deregulation of global telecommunications products and services continues, this model of national-security protection has manifest limitations. For example, there is virtually no guarantee that either the American or British security models will be able to detect the existence of compromised network switches with advanced interception capabilities. Such switches, usually designed and manufactured abroad, are used by the vast array of domestic and foreign telecommunications companies that are active in the United States and the United Kingdom.

Regardless of the long-term prospects of global capitalism, future research on communications interception should consistently engage with the issue of governmental regulation of telecommunications and its impact on information security. Undoubtedly, policy-makers who are in support of CALEA and RIPA view the concept of state-sponsored communications interception as vital for social regulation. In other words, both pieces of legislation are intended to safeguard interests that are perceived as socially necessary, but which the profit-oriented telecommunications market cannot be expected to meet voluntarily (Hills 1986, p. 29; Interview 19 2000, 220ff). Yet, the financial impact of CALEA and RIPA, especially on smaller carriers, cannot be overlooked. Interviews conducted in the context of the present research reveal a consistent concern among observers that small-sized carriers may be unable to withstand the financial requisites of wiretapping legislation:

[large TSPs] have such tremendous resources, whereas smaller companies don't have the resources [. . .]. CALEA requirements go] far beyond the capabilities of the small companies to [. . .] deal with this. All of the networking and the aggregation of [. . .] the data redistribution and the single point [. . .]—that's just [. . .] a level way above what these small companies that we represent would ever be capable of doing. So [. . .] there's surely a problem with small companies having this imposed on them for [. . .] no reason [Technical Director, national association representing over 500 smaller TSPs, Interview 17 2000, pp. 77–84];

I think [. . .] there will be more major [industry] players in the future, rather than small-scale TSPs. Th[e latter] will probably die in [. . .] light of these new developments, like the RIP Bill [. . .]. It's becoming a complex business to be [. . .] a carrier or run a carrier without [being] hassled by [. . .] law enforcement people [Akdeniz 2000, pp. 450–455; see also Interview 07 2000, pp. 61–62; Virgo 1999; Clayton et al. 2000].

These warnings, expressed nearly 20 years ago, were not unfounded. The financial impact of communications-interception regulations may have potentially contributed to the elimination of competition in the deregulated telecommunications environment. It may thus have gradually helped turn the market toward an oligopolistic direction—which, as explained earlier, could in turn black-box the communications interception debate. Scholarly research should therefore explore the state's intentions behind the future iterations of digital communications-interception legislation. In doing so, it should examine the extent to which CALEA and RIPA were partially used by elements within the American and British governments as deliberate instruments of policy in an orchestrated attempt to reduce the multitude of industry actors involved in the communications interception debate (Bowden 2000, pp. 678–684).

None of the above suggestions for further research can even begin to be addressed unless information-security experts systematically turn their attention toward communications interception and other forms of state-sponsored surveillance. It is true that, especially in its domestic form, surveillance has traditionally been a politically charged area for academic research. It is an uncomfortable subject-matter that becomes even more unpleasant when discussed in the context of parliamentary democracy. This, however, is hardly an adequate justification for the self-imposed distancing of information-security research from the theory and practice of state surveillance. This distancing has undermined attempts to adequately comprehend the nature and inner workings of governance in the West. During the Cold War, it helped create the illusion among many academics that the structure of Western communications systems was somehow unconstrained by institutional manipulation and control (Frey 1973; Almond 1960; Pye 1963). During the rise of digital networking, it contributed to popularized 'feel-good' analyses, produced with minimal reference to concepts of power and authority in technologically advanced social systems (Toffler 1980; Negroponte 1995).

And yet, the technological sophistication of the information society does not isolate it from the functional characteristics of preceding modes of social development: there still exist social classes, criminal actors, centralized governing bodies, wars, and mass movements. What is more, the networks of telecommunication through which power is negotiated on a daily basis are not neutral. They embody technical visions of the political order under which we live now or may live under in the future. Researchers cannot afford to ignore these visions any more than architects can ignore the density of the soil on which future buildings are to be erected. When this is fully understood, we can perhaps begin to deconstruct the present realities, as well as the possible futures, of the information society.

References

Ackroyd, C., et al. (1980) *The Technology of Political Control*, Penguin Books, Harmondsworth, Middlesex.

Ahmed, K. (2000) "Secret Plan to Spy on All British Phone Calls", *The Observer*, London, 03 December 2000

Aid, M.M. (2009) *The Secret Sentry: The Untold History of the National Security Agency*, Bloomsbury Press, New York, NY.

Akdeniz, Y. (2000) "New Privacy Concerns: ISPs, Crime Prevention and Consumers' Rights", *International Review of Law, Computers and Technology*, 14(1), pp. 55-61.

Akdeniz, Y., et al. (1999) *Who Watches the Watchmen Part III: ISP Capabilities for the Provision of Personal Information to the Police*, Cyber-Rights & Cyber Liberties UK, Leeds, February.

Almond, G.A. (1960) "A Functional Approach to Comparative Politics", in G.A. Almond and J.S. Coleman (eds.) *The Politics of the Developing Areas*, Princeton University Press, Princeton, NJ, pp. 94-128.

American Friends Service Committee (1979) *The Police Threat to Political Liberty*, American Friends Service Committee, Philadelphia, PA.

Anderson, R., et al. (1999) "Unprecedented Safeguards for Unprecedented Capabilities", paper presented at *Hoover Institution National Security Forum Conference* entitled *International Cooperation to Combat Cyber Attacks*, Stanford University, Stanford, CA, 07 December 1999 (copy on file with author).

Andrew, C. (1985) *Secret Service: The Making of the British Intelligence Community*, Heinemann, London.

Andrew, C., and Gordievsky, O. (1990) *KGB: The Inside Story of Its Foreign Operations from Lenin to Gorbachev*, Harper Collins, New York, NY.

Anonymous (1994) "British Telecom Security Contact Information", *2600 Magazine: The Hacker Quarterly*, New York, NY, summer.

Anonymous (1996) "Clinton Administration, Congress, Propose Sweeping Anti-Terrorism Initiatives", *CDT Policy Post*, 2(29), Center for Democracy and Technology, Washington, DC, 01 August.

Anonymous (1997a) *Industry and Privacy Advocates' Response to FBI CALEA Implementation Plan*, CTIA, USTA, PCIA, CDT, Washington, DC, 29 April, <http://www.cdt.org/digi_tele/970429_resp.html>.

Anonymous (1997b) *Analysis of Emerging Technologies and Services*, Booz Allen & Hamilton, Washington, DC, March (marked "for official use only", copy on file with author).

Anonymous (1998a) *FBI Tries to Use CALEA to Expand Its Surveillance Capabilities*, Center for Democracy and Technology, Washington, DC, <http://www.cdt.org/digitele/expand.html>.

© Springer Nature Switzerland AG 2020
J. Fitsanakis, *Redesigning Wiretapping*, History of Information Security,
https://doi.org/10.1007/978-3-030-39919-1

Anonymous (1998b) *Big Brother in the Wires: Wiretapping in the Digital Age*, American Civil Liberties Union, Wye Mills, Maryland, March, <https://www.aclu.org/other/big-brother-wires-wiretapping-digital-age>.

Anonymous (1998c) "FBI Claims Substantial Progress, Facts Show Otherwise", *CDT Policy Post*, 4(3.2), 18 February.

Anonymous (1999a) *Response to the Interception of Communications Consultation Paper*, Internet Service Providers Association, London, August <http://www.ispa.org.uk/html/august_1999.htm>.

Anonymous (1999b) *Response to the Interception of Communications Act Consultation Exercise*, Scottish Telecom, Glasgow, 13 August (copy on file with author).

Anonymous (1999c) *CALEA: A Precedent for Domestic Encryption Controls?*, Center for Democracy and Technology, Washington, DC.

Anonymous (1999d) *Interception of Communications in the United Kingdom*, Nortel Networks, London (copy on file with author).

Anonymous (1999e) *Alliance for Electronic Business Response to Interception of Communications in the UK*, Alliance for Electronic Business, London, August.

Anonymous (2000a) *STAND's Guide to the RIP v1.0*, STAND, London, 29 February, <http://www.stand.org.uk/ripnotes>.

Anonymous (2000b) "Law Enforcement Issues", presentation given at closed-door industry meeting, London, England, 12 April (copy on file with author).

Appleby, C. (1995) "Making Security a Reality for All", *Information Week*, New York, NY, 09 January 1995, <http://www.infoweek.com/509/08encry.htm>.

Apter, D.E. (1965) *The Politics of Modernisation*, University of Chicago Press, Chicago, IL.

Aronson, S.H. (1977) "Bell's Electrical Toy: What's the Use? The Sociology of Early Telephone Usage", in I. de S. Pool (ed.) *The Impact of the Telephone*, MIT Press, Cambridge, Massachusetts, pp. 15-39.

Auble, D., and Bender, D. (2000) "CALEA Capacity Requirements: What You Need To Know", presentation given at the *103rd Annual United States Telecommunications Association Convention and Exhibition*, Miami Beach, Florida, 02 October.

Baker, S.A (1994) "Don't Worry Be Happy", *Wired Magazine*, San Francisco, CA, issue 2.06, June.

Baker, S.A., et al. (1998) *Comments of the Telecommunications Industry Association Before the Federal Communications Commission in the Matter of CALEA*, CC Docket No 97-213, Washington, DC, 08 May (draft version).

Baleanu, V.G. (1995) *The Enemy Within: The Romanian Intelligence Service in Transition*, Royal Military Academy Conflict Studies Research Centre, Camberley, Surrey.

Ball, D. (1989a) "Soviet Signals Intelligence: Vehicular Systems and Operations", *Intelligence and National Security*, 4(1), January, pp. 29-44.

Ball, D. (1989b) *Soviet Signals Intelligence SIGIN77: Intercepting Satellite Communications*, Australian National University Strategic and Defence Studies Centre, Canberra.

Barlow, J.P. (1994) "Jackboots on the Infobahn", *Wired Magazine*, issue 2.04, San Francisco, February, <http://www.cpsr.org/cpsr/privacy/crypto/clipper/barlow_wired_2.04_clipper>.

Barlow, J.P., and Denning, D.E. (1994) "Online Debate Over the Clipper Chip Scheme", *Time Online forum of America On Line*, 10 March <https://www.eff.org/pages/barlow-denning-clipper-chip-scheme>.

Barry, E.E. (1965) *Nationalisation in British Politics: The Historical Background*, Jonathan Cape, London.

Baxter, J. D. (1990) *State Security, Privacy and Information*, Harvester & Wheatsheaf, New York, NY.

Beer, M. (1920) *A History of British Socialism*, vol. II, G. Bell & Sons, London.

Berman, J., and Dempsey, J.X. (1996) *Comments on CALEA Capacity Notice, 60 Fed. Reg. 53643*, letter addressed to David F. Worthley, Unit Chief, Federal Bureau of Investigation Telecommunications Interception Liaison Unit, Washington, DC, 16 January.

Bertrand Russell Peace Foundation (1972) *Subversion in Chile: A Case Study in US Corporate Intrigue in the Third World*, Spokesman Books, Nottingham.

Blacharski, D. (1998) *Network Security in a Mixed Environment*, IDG Books, New York, NY.

Blair, P.D., et al. (1995) *Electronic Surveillance in a Digital Age*, US Congress Office of Technology Assessment, Government Printing Office, Washington, DC, July.

Blank, S. (1973) *Industry and Government in Britain: The Federation of British Industries in Politics, 1945-1965*, Saxon House, Farnborough, Hants.

Bleep, B. (1992) "BT's Fight Against Fraud", *BT Today Magazine*, London, June.

Bloss, W. (2007) "Escalating US Police Surveillance after 9/11: An Examination of Causes and Effects", *Surveillance and Criminal Justice*, 4(3), pp. 208-228.

Bolling, G.H. (1983) *AT&T, Aftermath of Antitrust: Preserving Positive Command and Control*, National Defense University, Washington, DC.

Boucher, R., and Edwards, D. (1994) *Telecommunications Carrier Assistance to the Government*, Report 103-827, 103rd Congress, 2nd Session, FBI, Washington, DC, 04 October.

Bourne, I. (2001) "Data Retention and Law Enforcement Requirements: How Can the Two be Reconciled?", presentation given at the *12th Internet Service Providers Association Legal Forum*, London, 05 July.

Bowden, C., (2000) Director, *Foundation for Information Policy Research*, Interview to Joseph Fitsanakis, London, England, Wednesday 12 April 2000 (recorded).

Brenton, M. (1964) *The Privacy Invaders*, Coward-McCann Inc., New York, NY.

Briggs, A. (1961) *The History of Broadcasting in the United Kingdom*, vol. I, Oxford University Press, Oxford.

Brooke, S. (1991) "Atlantic Crossing? American Views of Capitalism and British Socialist Thought, 1932-1962", *Twentieth Century British History*, 2(2), pp. 107-136.

Brown, D. (2001) "Big Brother to Test Tapping Capabilities", *Interactive Week*, San Francisco, CA, 18 June.

Brown, I., et al. (eds.) (2000) *The Economic Impact of the Regulation of Investigatory Powers Bill*, British Chambers of Commerce, London, 18 August.

Bunyan, T. (1976) *The History and Practice of the Political Police in Britain*, Julian Friedmann Publishers, London 1976.

Cadwalladr, C., and Graham-Harrison, E. (2018) "Facebook Accused of Conducting Mass Surveillance Through its Apps", *The Guardian*, 24 May.

Caiden, G.E. (1977) *Police Revitalisation*, Lexington Books, Lexington, MA.

Canes, M. (1966) *Telephones: Public or Private? A Comparative Study of the British and American Systems*, Institute of Economic Affairs, London.

Carr, J.G. (1998) *Law on Electronic Surveillance*, West Group, St. Paul, MN, 2nd edition, release No 23.

Carroll, J.A. (1969) *The Third Listener*, E.P. Dutton, New York, NY.

Carter, A.B. (1989) "Telecommunications Policy and US National Security", in R.W. Crandall and K. Flamm (eds.) *Changing the Rules: Technological Change, International Competition and Regulation in Communications*, The Brookings Institution, Washington, DC, pp. 221-256.

Castells, M. (1989) *The Informational City: Information Technology, Economic Restructuring, and the Urban-Regional Process*, Basil Blackwell, Oxford.

Charlesworth, A., et al. (1996) *An Atlas of Industrial Protest in Britain, 1750-1990*, Macmillan, Basingstoke, Hampshire.

Charney, S., and Alexander, K. (1996) "Computer Crime", paper presented at the *Randolph W. Thrower Symposium: Legal Issues in Cyberspace: Hazards on the Information Superhighway*, 26-28 February, Atlanta, Georgia.

Chick, M. (1991) "Competition, Competitiveness and Nationalisation, 1945-1951", in G. Jones and M.W. Kirby (eds.) *Competitiveness and the State: Government and Business in Twentieth-Century Britain*, Manchester University Press, Manchester, pp. 60-77.

Cisneros, O.S. (2000) "These Wires Were Made for Tapping", *Wired News*, San Francisco, CA, 14 August, <http://www.wired.com/news/politics/0,1283,38170, 00.html>.

Clark, J., et al. (1989) *The Process of Technological Change: New Technology and Social Choice in the Workplace*, Cambridge University Press, Cambridge.

Clayton, R., et al. (2000) *Response to the Smith Group Report for the Home Office on Technical and Cost Issues Associated with Interception of the Internet*, Internet Service Providers Association, London, 23 May, <http://www.ispa.org.uk/html//updateable_smithreport.htm>.

Cohen, J.E. (1992) *The Politics of Telecommunications Regulation: The States and the Divestiture of AT&T*, M.E. Sharpe, Armonk, NY.

Cohn, C. (2000) *CALEA Impact*, electronic message addressed to Joseph Fitsanakis, Electronic Frontier Foundation, San Francisco, CA, 29 June (copy on file with author).

Cole, C. (1999a) *Internet Users Privacy Forum Minutes: Meeting One*, London Internet Exchange, Peterborough, 22 March.

Cole, C. (1999b) *Internet Users Privacy Forum Minutes: Meeting Two*, Easynet, London, 02 June.

Command 108 (1987) *Interception of Communications Act*, Her Majesty's Stationery Office, London, March.

Command 283 (1957) *Report of the Committee of Privy Councillors Appointed to Inquire into the Interception of Communications*, Her Majesty's Stationery Office, London, September [often referred to as the 'report of the *Birkett Commission*'].

Command 4778 (2000) *Interception of Communications Act 1985: Report of the Commissioner for 1999*, Her Majesty's Stationery Office, London, July.

Command 7873 (1980) *The Interception of Communications in Great Britain*, Her Majesty's Stationery Office, London, April.

Command 9843 (1985) *The Interception of Communications in Great Britain*, Her Majesty's Stationery Office, London, February.

Corn-Revere, R. (2000) *The Fourth Amendment and the Internet: Testimony Before the Subcommittee on the Constitution of the Committee on the Judiciary*, US House of Representatives, Washington, DC, 06 April.

Courtney, R. (2000) *CDT and CALEA*, electronic message addressed to Joseph Fitsanakis, Center for Democracy and Technology, Washington, DC, 03 July (copy on file with author).

Cruise-O'Brien, R., and Helleiner, G.K. (1983) "The Political Economy of Information in a Changing International Economic Order", in R.C. Cruise-O'Brien (ed.) *Information, Economics and Power: The North-South Dimension*, Hodder & Stoughton, London, pp. 1-27.

Curry, J.C. (1999) *The Security Service 1908-1945: The Official History*, Public Record Office, Kew.

Cutright, P. (1963) "National Political Development: Measurement and Analysis", *American Sociological Review*, 28(2), April, pp. 253-264.

Dandeker, C. (1990) *Surveillance, Power and Modernity: Bureaucracy and Discipline from 1700 to the Present Day*, Polity Press, Cambridge.

Danelian, N.R. (1939) *AT&T: The Story of Industrial Conquest*, Vanguard Press, New York, NY.

Dash, S., et al. (1959) *The Eavesdroppers*, Rutgers University Press, New Brunswick, NJ.

de Sola Pool, I., et al. (1977) "Foresight and Hindsight: The Case of the Telephone", in I. de S. Pool (ed.) *The Impact of the Telephone*, MIT Press, Cambridge, MA, pp. 127-157.

de Stempel, C. (2001) "AOL's monitoring costs", comments made at the *12th Internet Service Providers Association Legal Forum*, London, 05 July.

Delio, M. (2001) "ISPs 'RIP' Into British Police", *Wired News*, San Francisco, CA, 19 January, <http://www.wired.com/news/politics/0,1283,41288,00.html>.

Demac, D.A. (1988) "Hearts and Minds Revisited: The Information Policies of the Reagan Administration", in V. Mosco and J. Wasko (eds.) *The Political Economy of Information*, The University of Wisconsin Press, Madison, WI, pp. 125-145.

Dempsey, J.X. (1995) *Memorandum on Capacity Requirements Under CALEA*, Centre for National Security Studies, Washington, DC, 09 November.

Dempsey, J.X. (1998) *Comments Before the Federal Communications Commission in the Matter of CALEA Extension of October 1998 Compliance Date*, CC Docket No 97-213, Center for Democracy and Technology, Washington, DC, 02 April.

Dempsey, J.X. (1999) *FBI Seeks to Impose Surveillance Mandates on Telephone System; Balanced Objectives of 1994 Law Frustrated*, Center for Democracy and Technology, Washington, DC, 04 March.

Dempsey, J.X., and Oram, A. (1998) *Reply Comments Before the Federal Communications Commission Regarding the Implementation of CALEA*, CC Docket No 97-213, Washington, DC, 11 February.

Denning, D.E. (1993) "To Tap or Not to Tap", *Communications of the ACM*, 36(3), pp. 377-383.

Denning, D.E. (1994) "Campaign and Petition Against Clipper", *Privacy Forum Digest*, 3(4), Woodland Hills, CA, 20 February, pp. 566-575.

Department of Trade and Industry (1996) *Development of the Information Society: An International Analysis*, Her Majesty's Stationery Office, Norwich.

Deutsch, K.W. (1963) *The Nerves of Government: Models of Political Communication and Control*, The Free Press, New York City, NY.

Diffie, W., and Landau, S. (1998) *Privacy on the Line: The Politics of Wiretapping and Encryption*, MIT Press, Cambridge, MA.

Dillon, M. (1999) *The Dirty War: Covert Strategies and Tactics Used in Political Conflicts*, Routledge, New York, NY.

Dilts, M.M. (1914) *The Telephone in a Changing World*, Longman's Green, New York, NY.

Dodd, Vikram (2009) "Ministers to review councils' use of anti-terrorism surveillance powers", *The Guardian*, 17 April.

Donner, F.J. (1980) *The Age of Surveillance: The Aims and Methods of America's Political Intelligence System*, Alfred A. Knopf, New York, NY.

Dordick, H.S., and Wang, G. (1993) *The Information Society: A Retrospective View*, Sage Publications, London.

Douglas, W.O. (1954) *An Almanac of Liberty*, Doubleday, Garden City, NY.

Downes, L. (2014) "The End of the Line for the Analog Phone Network" *Harvard Business Review*, 28 March.

EITC (1868) *Government and Telegraphs: Statement of the Case of the EITC Against the Government Bill for Acquiring the Telegraphs*, Electric and International Telegraph Company, London.

Electronic Surveillance Task Force (1997) *Communications Privacy in the Digital Age: Interim Report*, Digital Privacy and Security Working Group, Washington, DC, June, <http://www.cdt.org/digi_tele/9706rpt.html>.

Elliff, J.T. (1972) "The FBI and Domestic Intelligence", in R.H. Blum (ed.) Surveillance and Espionage in a Free Society: A Report by the Planning Group on Intelligence and Security to the Policy Council of the Democratic National Committee, Praeger Publishers, New York, NY, pp. 20-45.

Ellul, J. (1965) *The Technological Society*, Johnathan Cape, London.

Ellul, J. (1989) *What I Believe*, Eerdmans, London.

Emery, D. (1999) *Consultation Response*, International Computers Limited, Slough, 10 August (copy on file with author).

Emsley, C. (1996) *The English Police: A Political and Social History*, Longman, London, 2nd edition.

Ericson, R.V., and Haggerty, K.D. (1997) *Policing the Risk Society*, University of Toronto Press, Toronto.

Etzioni-Halevy, A. (1983) *Bureaucracy and Democracy: A Political Dilemma*, Routledge & Kegan Paul, London.

European Commission (1995) *G7 Ministerial Conference on the Global Information Society*, Office for Official Publications of the European Communities, Luxembourg.

Evans, V. (1999) *Internet Users Privacy Forum Minutes: Meeting Three*, Easynet, London, 02 September.

Evans, E.J. (2004) *Thatcher and Thatcherism*, Routledge, London, 2nd edition.

Feldman, J. (2000) "Digital Sculduggery: Online Sleuths Tell All", transcript of Interview with Dana Hawkins, *US News and World Report*, 02 October 2000, <http://www.usnews.com/usnews/issue/001002/nycu/privacy.b.htm>.

Fitzgerald, P. (1994) "All About Eavesdropping", *The New Statesman & Society*, London, 29 July, pp. 30-31.

Fitzgerald, P., and Leopold, M. (1987) *Stranger on the Line: The Secret History of Phone Tapping*, The Bodley Head, London.

Flaherty, D. (1989) *Protecting Privacy in Surveillance Societies: The Federal Republic of Germany, Sweden, France, Canada and the United States*, University of North Carolina Press, Raleigh, NC.

Foucault, M. (1977) *Discipline and Punish: The Birth of the Prison*, Pantheon Books, New York City, NY.

Freeh, L. (1995) Letter addressed to Henry Hyde, Chairman of the US House of Representatives Committee on the Judiciary, Federal Bureau of Investigation, Chantilly, VA, 02 November, <http://www.cdt.org/policy/digtel/freeh_hyde_ltr.html>.

Freeh, L. (1997) *Excerpts from Testimony to the Senate Judiciary Committee*, Center for Democracy and Technology, Washington, DC, 04 June, <http://www.ctd.org/digi_tele/970604_Freeh.html>.

Freeh, L. (2000) *Statement Before the Subcommittee on Technology, Terrorism and Government Information of the US Senate Committee on the Judiciary on Examining the Incidence of Cyber Attacks on the Nation's Information Systems: Removing Roadblocks to Investigation and Information Sharing*, S. Hrg. 106-839, 106th Congress, 2nd Session, Government Printing Office, Washington, DC, 28 March.

Frey, F.W. (1973) "Communication and Development", in I. de S. Pool and W. Schramm (eds.) *Handbook of Communication*, Rand McNally, Chicago, IL, pp. 337-461.

Friedman, L. (1973) *A History of American Law*, Simon & Schuster, New York, NY.

Fuld, L. (1909) *Police Administration*, Patterson Smith, New York, NY.

Galbraith, J.K. (1967) *The New Industrial State*, Houghton Mifflin Co., Boston, MA.

Gandy, Jr., O.H. (1989) "The Surveillance Society: Information Technology and Bureaucratic Social Control", in M. Siefert, G. Gerbner & J. Fisher (eds.) *The Information Gap: How Computers and Other New Communication Technologies Affect the Social Distribution of Power*, Oxford University Press, Oxford, pp. 61-76.

Gandy, Jr., O.H. (1993) *The Panoptic Sort: A Political Economy of Personal Information*, Westview Press, Boulder, CO.

Garnet, R.W. (1985) *The Telephone Enterprise: The Evolution of the Bell System's Horizontal Structure, 1876-1909*, Johns Hopkins University Press, Baltimore, MD.

Gaspar, R. (2000) *Looking to the Future: Clarity of Communications Data Retention Law*, National Criminal Intelligence Service Submission on Communications Data Retention Law, London 21 August (copy on file with author).

Giddens, A. (1985) *The Nation State and Violence: A Contemporary Critique of Historical Materialism*, Polity Press, Cambridge.

Gill, P. (1994) *Policing Politics: Security Intelligence and the Liberal Democratic State*, Frank Cass, London.

Goralski, W.J., and Kolon, M. (2000) *IP Telephony*, McGraw-Hill, San Francisco, CA.

Grey, D.L. (1999) *Consultation Paper on the Interception of Communications In the United Kingdom*, letter addressed to the Home Office Interception Legislation Team, Durham Police Constabulary, Durham, 15 July (copy on file with author).

Hall, Kat (2018) "BT Pushes Ahead With Plans to Switch Off Telephone Network", *The Register*, 19 April.

Hall, C., et al. (1999) *Response to the Interception of Communications Act Consultation Exercise*, Demon Internet, London, 19 August (copy on file with author).

Harden, I., and Lewis, N. (1988) *The Noble Lie: The British Constitution and the Rule of Law*, Hutchinson, London.

Hayden, M., Dershowitz, A., Greenwald, G., and Ohanian, A. (2014) *Does State Spying Make Us Safer?: The Munk Debate on Mass Surveillance*, The House of Anansi, Toronto, Canada.

Hayward, S.F. (2009) *The Age of Reagan: The Conservative Counterrevolution, 1980-1989*, Three Rivers Press, New York, NY.

HCSCF (2000) *Regulation of Investigatory Powers Bill: First Sitting*, Report of Proceedings, House of Commons Standing Committee F, London, 14 March.

HCTISC (1999a) *Seventh Report: Building Confidence in Electronic Commerce*, HC 187, House of Commons Trade and Industry Select Committee, Her Majesty's Stationery Office, London, 19 May.

HCTISC (1999b) *Fourteenth Report: Draft Electronic Communications Bill*, HC 862, House of Commons Trade and Industry Select Committee, Her Majesty's Stationery Office, London, November.

HCTISC (2000) *Third Special Report: Further Government Observations on the Fourteenth Report from the Trade and Industry Committee on the Draft Electronic Communications Bill*, HC 199, House of Commons Trade and Industry Select Committee, Her Majesty's Stationery Office, London, February.

Headrick, D.R. (1991) *The Invisible Weapon: Telecommunications and International Politics, 1851-1945*, Oxford University Press, New York.

Heap, N., et al. (1995) *Information Technology and Society: A Reader*, Open University Press, London.

Herrmann, D.G. (1996) *The Arming of Europe and the Making of the First World War*, Princeton University Press, Princeton, NJ.

Heupel, R. (1999) "CommNet Cellular Signs US$35 Million Deal with Nortel Networks For Network Modernization Software Upgrades", *Nortel Networks News Release*, Brampton, Ontario, 19 January, <http://www.notelnetworks.com/corporate/news/newsreleases/1999a/1_19_9999016_CommNet.html>.

Hildebrand, T.A. (2000) "TMIC Notes", *This Month in CALEA*, FBI, CIS, Raleigh, NC, 14 November.

Hills, J. (1986) *Deregulating Telecoms: Competition and Control in the United States, Japan and Britain*, Frances Pinter, London.

Hinton, J. (1983) *Labour and Socialism: A History of the British Labour Movement, 1867-1974*, Wheatsheaf Books, Brighton, Essex.

Home Office (1977) *Consolidated Circular to the Police on Crime and Kindred Matters*, Her Majesty's Stationery Office, London.

Home Office (2001a) Regulation of Investigatory Powers Act 2000: Consultation on the Section 13 Order, London, 09 March, <http://www.homeoffice.gov.uk/ripa/consec13.htm>.

Home Office (2001b) *Consultation on Section 12 of the Regulation of Investigatory Powers Act 2000*, London, 29 June, <http://www.homeoffice.gov.uk/ripa/section12.htm>.

Home Office (2001c) *National Technical Assistance Centre*, London 15 August, <http://www.homeoffice.gov.uk/oicd/ecu/ntac.htm>.

Home Office (2001d) *Head of National Technical Assistance Centre (NTAC) Announced*, Home Office News Release, London, 30 March 2001.

Horner, M. (1996) *Information, Technology and the Information Society*, The British Library, London.

House of Commons (2000) *Regulation of Investigatory Powers Bill*, Hansard vol. 345, cols 776-837, 02 November.

Hughes, P. (1981) *British Broadcasting: Programmes and Power*, Chartwell-Bratt, Bromley, Kent.

Hunt, E.H. (1981) *British Labour History, 1815-1914*, Weidenfeld & Nicolson, London.

Interview 01. Director, CALEA Compliance Unit, large national telecommunications service provider, Miami Beach, Florida, 01 October 2000 (recorded).

Interview 02. Communications Intelligence Officer, The Security Service (MI5), London, United Kingdom, 12 April 2000 (recorded).

Interview 03. External Maintenance Engineer (former), Engineering Division, British Telecom, Plc, Bristol, United Kingdom, 10 November 2001 (recorded).

Interview 04. Fraud Control and Revenue Assurance Manager, large British wireless communications service provider, Reading, United Kingdom, 09 May 2000 (recorded).

Interview 05. Officer, National Security Agency, Baltimore, Maryland, 11 December 2000 (recorded).

Interview 06. Fraud Control Manager, large British telecommunications service provider, Reading, United Kingdom, 12 May 2000 (recorded).

Interview 07. Security Officer, large British telecommunications service provider, London, United Kingdom, 08 May 2000 (recorded).

Interview 08. Chief Executive Officer, large British Internet service provider, Paisley, United Kingdom, 28 April 2000 (recorded).

Interview 09. Fraud Liaison Officer, large British TSP telecommunications service provider, London, 03 May 2000 (recorded).

Interview 10. Regulation Officer, large British Internet service provider association, Peterborough, United Kingdom, 10 May 2000 (recorded).

Interview 11. Regulatory Affairs Officer, large British TSP telecommunications service provider, London, United Kingdom, 12 April 2000 (recorded).

Interview 12. Officer, The Security Service (MI5), London, United Kingdom, 11 April 2000 (recorded).

Interview 13. Technical Unit Chief, Emergency Communications Unit, state government agency, Bluff City, Tennessee, 23 December 2000 (recorded).

Interview 14. Police Officer with experience in undercover and wiretapping operations, United States, Franklin, Tennessee, 12 January 2001 (recorded).

Interview 15. Sergeant, large United States police department, New York, NY, 02 June 2000 (recorded).

Interview 16. CALEA Lobbyist, large telecommunications service provider association, Miami Beach, Florida, 03 October 2000 (recorded).

Interview 17. Technical Director, national association representing over 500 smaller TSPs telecommunications service providers, Miami Beach, Florida, Monday 02 October 2000 (recorded).

Interview 18. Government Relations Director, large national telecommunications service provider association, Miami Beach, Florida, 02 October 2000 (recorded).

Interview 19. Special Agent and Unit Chief, Telecommunications Industry Liaison Unit, CALEA Implementation Section, Federal Bureau of Investigation, Atlanta, Georgia, 15 November 2000 (recorded).

Interview 20. CALEA Compliance Attorney, large United States telecommunications service provider, Atlanta, Georgia, 15 June 2001 (recorded).

Interview 21. CALEA Project Manager, large United States TSP telecommunications service provider, Atlanta, Georgia, 15 June 2001 (recorded).

Interview 22. Officer, National Criminal Intelligence Service, London, United Kingdom, 01 April 2001 (recorded).

Interview 23. Law Enforcement Liaison, large British telecommunications service provider, Crewe, United Kingdom, 15 November 2001 (recorded).

Interview 24. Fraud Specialist, large British telephone service provider, Kent, United Kingdom, 09 September 2001 (recorded).

Interview 25. Technical Specialist, large British telecommunications service provider, London, United Kingdom, 08 October 2001 (recorded).

Interview 26. Chief Operating Officer, large British Internet service provider, Edinburgh, United Kingdom, 1 April 2001 (recorded).

Interview 27. Security Officer, large British Internet service provider, Edinburgh, United Kingdom, 4 April 2001 (recorded).

Interview 28. Chief, Fraud Department, large British telecommunications service provider, St. Andrews, 08 April 2001 (recorded).

Interview 29. Law Enforcement Liaison, large British Internet service provider association, London, United Kingdom, 28 March 2001 (recorded).

Interview 30. Law Enforcement Liaison and Fraud Specialist, large British Internet service provider association, Glasgow, United Kingdom, Wednesday 3 April 2001 (recorded).

Interview 31. Regulation Officer, large British TSP telecommunications service provider, London, United Kingdom, 09 October 2001 (recorded).

Interview 32. Head, Regulation Affairs Department, large British TSP telecommunications service provider, London, United Kingdom, 10 October 2001 (recorded).

Interview 33. Public Affairs Officer, The Security Service (MI5), London, United Kingdom, 12 April 2000 (recorded).

Interview 34. Head, Emergency Communications Unit, state government agency, Bluff City, Tennessee, 23 December 2000 (recorded).

Interview 35. FBI Special Agent in Charge, United States, Atlanta, Georgia, 11 June 2001 (recorded).

Interview 36. Detective, large United States police department, Newark, New Jersey, 01 June 2000 (recorded).

Interview 37. Director, National Services Planning, large telecommunications service provider association, Miami Beach, Florida, 02 October 2000 (recorded).

Interview 38. Policy Lobbyist, national association of rural telecommunications service providers, Miami Beach, Florida, Monday 02 October 2000 (recorded).

Interview 39. CALEA Director, large national telecommunications service provider association, Miami Beach, Florida, 02 October 2000 (recorded).

Interview 40. Special Agent and Telecommunications Industry Liaison, CALEA Implementation Section, Federal Bureau of Investigation, Atlanta, Georgia, 14 November 2000 (recorded).

Interview 41. CALEA Administrator, large United States telecommunications service provider, Atlanta, Georgia, Friday 15 June 2001 (recorded).

Iosifidis, P. (2014) "United Kingdom", in E. Noam (ed.) *Who Owns the World's Media? Media Concentration and Ownership Around the World*, Oxford University Press, New York, NY, pp. 425-453

Janson, M.A., and Yoo, C.S. (2013) *The Wires Go to War: The US Experiment with Government Ownership of the Telephone System During World War I*, Faculty Scholarship Paper, no. 467, The University of Pennsylvania Law School, Philadelphia, PA.

Jeffery, K., and Hennessy, P. (1983) *States of Emergency: British Governments and Strikebreaking Since 1919*, Routledge & Kegan Paul, London.

Jeffreys, D. (1995) *The Bureau: Inside the Modern FBI*, Houghton Mifflin Company, New York, NY.

Jepperson, R.L., et al. (1996) "Norms, Identity and Culture in National Security", in P.J. Katzenstein (ed.) *The Culture of National Security: Norms and Identity in World Politics*, Columbia University Press, New York, NY, pp. 33-74.

Jiang, Y., Tonami, A., and Fejerksov, A.M. (2016) *China's Overseas Investment in Critical Infrastructure: Nuclear Power and Communications*, Danish Institute for International Studies, Copenhagen, Denmark.

Johnson, J. (1998) *The Evolution of British SIGINT, 1653-1939*, HM Stationery Office, London.

Joy, D., and Wright, R.O. (1974) "Wiretaps in Perspective", in R.O. Right (ed.) *Whose FBI?*, Open Court, La Salle, IL, pp. 251-265.

Judge, J.F. (1985) "The Defence Communications Agency: Coping with Changing Times", *Defence Electronics*, vol. XVII, May, pp. 230-236.

Kallstrom, J.K. (1997) *Statement Concerning the Second Notice of Capacity*, Federal Bureau of Investigation National Press Office, Washington, DC, 14 January.

Kampschoer, G. (2000) "RIP Law is the Least of It", *Professional Security Magazine*, Wolverhampton, 11 October.

Kern, S. (1983) *The Culture of Time and Space, 1880-1918*, Harvard University Press, Cambridge, MA.

Kerr, D.M. (2000) *Statement of the FBI Laboratory Division Before the USCJ: Carnivore Diagnostic Tool*, Washington, DC, 06 September.

Kessler, R. (1993) *The FBI: Inside the World's Most Powerful Law-Enforcement Agency*, Pocket Books, New York, NY.

Kieve, J. (1973) *The Electric Telegraph: A Social and Economic History*, David & Charles, Newton Abbot.

King, A. (1984) *The Coming Information Society*, The British Library, London.

Kirk, N. (1994) *Labour and Society in Britain and the USA: Challenge and Accommodation, 1850-1939*, vol. II, Scholar Press, Aldershot.

Knight, A. (1990) *The KGB: Police and Politics in the Soviet Union*, Unwin Hyman, Boston, MA.

Krasner, S.D (1978) *Defending the National Interest: Raw Materials Investments and US Foreign Policy*, Princeton University Press, Princeton, NJ.

Lambert, J.L. (1986) *Police Powers and Accountability*, Croom Helm, London.

Landau, S. (2010) *Surveillance or Security? The Risks Posed by New Wiretapping Technologies*, The MIT Press, Boston, MA.

Lansman, N. (2000) Letter addressed to Charles Clarke, MP, Minister of State, Internet Service Providers Association, London, 31 May, <http://www.ispa.org.uk/html//updateable_ccletter.htm>.

Lapidus, E.J. (1974) *Eavesdropping on Trial*, Hayden Book Company, Rochelle Park, NJ.

Leahy, P. (1995) Letter addressed to Louis Freeh, Director of the FBI, United States Congress, Washington, DC, 03 November, <http://www.cdt.org/policy/digtel/leahy_freeh_ltr.html>.

Leahy, P. (1999) *Statement Before the US Senate Committee on the Judiciary, Subcommittee on Criminal Justice Oversight During Hearing on the Responsibilities and Activities of the Criminal Division of the Department of Justice*, S. Hrg. 106-725, 106th Congress, 1st Session, United States Government Printing Office, Washington DC, 17 July.

Lean, T. (2016) *Electronic Dreams: How 1980s Britain Learned to Love the Computer*, Bloomsbury Publishing, London.

Lefort, C. (1986) *The Political Forms of Modern Society: Bureaucracy, Democracy, Totalitarianism*, Polity Press, Cambridge.

Leigh, L.H. (1975) *Police Powers in England and Wales*, Butterworth, London.

LeMond, A., and Fry, R. (1975) *No Place to Hide: A Guide to Bugs, Wire Taps, Surveillance and other Privacy Invasions*, St Martin's Press, New York City, NY.

Lemos, R. (2000) "FBI Releases First Carnivore Data", *ZDNet News*, San Francisco, CA, 04 October, <http://news.zdnet.co.uk/zdnetuk/news/story/0,s2081770,00.html>.

Leonard, V.A. (1938) *Police Communication Systems*, University of California Press, Berkeley, CA.

Leonard, V.A. (1964) *Police Organisation and Management*, The Foundation Press, Brooklyn, New York, 2nd edition.

Lethin, R. (1995) "Talk by Admiral Bobby Inman at MIT", *Computer Professionals for Social Responsibility*, Washington DC, February, <http://www.cpsr.org/cpsr/lists/rre/Bobby_Inman_on_Clipper_and_the>.

Levi-Faur, D. (2002) "New Regimes, New Capacities: The Politics of Telecommunications Nationalisation and Liberalisation", in Linda Weiss (ed.) *States in the Global Economy: Bringing Domestic Institutions Back In*, Cambridge University Press, Cambridge, UK, pp. 161-178.

Lipset, S.M. (1959) "Some Social Requisites of Democracy: Economic Development and Political Legitimacy", *American Political Science Review*, 53(3), March, pp. 69-105.

Lloyd, C. (1993) "Spymasters Order Redesign of 'Too Secure' Mobile Phones", *The Sunday Times*, 31 January.

Lustgarten, L., and Leigh, I. (1994) *In from the Cold: National Security and Parliamentary Democracy*, Clarendon Press, Oxford.

Lyon, D. (1994) *The Electronic Eye: The Rise of Surveillance Society*, Polity Press, Cambridge.

Macrakis, C. (2010) "Technophilic Hubris and Espionage Styles During the Cold War", *Isis*, **101** (2): 278-385.

Maddison, P.F. (1999) Letter addressed to the Home Office Organised and International Crime Directorate, Hertfordshire Constabulary, Welwyn Garden City, 23 July (copy on file with author).

Maguire, J.M. (1959) *Evidence of Guilt: Restrictions Upon its Discovery or Compulsory Disclosure*, Little, Brown & Company, Boston, MA.

Mansell, R. (1993) *The New Telecommunications: A Political Economy of Network Evolution*, Sage Publications, London.

Manwaring-White, S. (1983) *The Policing Revolution: Police Technology, Democracy and Liberty in Britain*, The Harvester Press, Sussex.

Marino, M.S. (2007) "The Problems of the First Mover: The Case of British Telecom", in K.A. Eliassen and J. From (eds.) *The Privatisation of European Telecommunications*, Ashgate, Aldershot, pp. 113-128.

Martin, M. (1991) *"Hello, Central?": Gender, Technology and Culture in the Formation of Telephone Systems*, McGill-Queen's University Press, Montreal.

Marvin, C. (1988) *When Old Technologies Were New: Thinking About Electric Communication in the Late Nineteenth Century*, Oxford University Press, New York, NY.

Masuda, Y. (1981) *The Information Society as Post-Industrial Society*, Institute for the Information Society, Tokyo.

Mavor, J. (1916) *Government Telephones: The Experience of Manitoba, Canada*, Moffat, Yard & Co., New York, NY.

McCullagh, D. (1998) "Shadow Cryptocrats", *CyberWire Dispatch*, Washington, DC, 04 February, <http://www.cyberwerks.com/cyberwire/cwd/cwd.98.02.25.html>.

McHugh, J., and Cordwell, B. (2000) "Carriers Must Comply With CALEA by May 2", *The OPASTCO Advocate*, Washington, DC, February, pp. 5-6.

McLure, J. (1984) *Cop World: Policing the Streets of San Diego*, Macmillan, London and Basingstoke.

Meeks, B.N. (1994) "The End of Privacy", *Wired Magazine*, issue 2.04, April.

Meeks, B.N. (2000) "Carnivore: The Truth is Worse Than You Thought", *ZDNet News*, San Francisco, CA, 18 October, <http://news.zdnet.co.uk/story/0,,t269s2082045,00.html>.

Miceli, J. (1998) *USTA and Other Carrier Organisations File Petition with FCC to Resolve Wiretap Issues*, USTA press statement, Washington, DC, 09 April, <http://www.cdt.org/rls98-21.html>.

Middlemas, K. (1986) *Power, Competition and the State. Volume I: Britain in Search of Balance, 1940-1961*, Macmillan, London.

Milano, J.V. (1995) *Soldiers, Spies and the Rat Line: America's Undeclared War Against the Soviets*, Brassey's Inc., Washington, DC.

Miles, I. (1988) *Information Technology and Information Society: Options for the Future*, Programme on Information and Communication Technologies Policy Research Paper, University of Sussex, Sussex.

Mills, E. (1998) "Privacy Advocates Snub 'Operation-Action' Encryption Plan", *The Industry Standard*, San Francisco, CA, 15 July <http://thestandard.net.articles/article_print/1,1454,1094,00.html>.

Minnick, W. (1995) "China Under Cover", *Far Eastern Economic Review*, 02 March, p. 38.

Molina, A.H. (1990) *The Development of Public Switching: Systems in the UK and Sweden*, Edinburgh Programme on Information and Communication Technologies Working Paper No.19, Edinburgh.

Morgan, J. (1987) *Conflict and Order: The Police and Labour Disputes in England and Wales, 1900-1939*, Clarendon Press, Oxford.

Morris, C.G. (1998) *Reply Comments of the FBI Before the Federal Communications Commission Regarding the Implementation of CALEA*, CC Docket No 97-213, Washington, DC, 11 February.

Mosquera, M. (1998) "Encryption Plan Lets United States Compete", *Techweb*, Manhasset, NY, 13 July, <http://www.techweb.com/wire/story/TWB19980713S0014>.

Mouzelis, N.P. (1967) *Organization and Bureaucracy: An Analysis of Modern Theories*, Routledge & Kegan Paul, London.

Munting, R. (1996) *An Economic and Social History of Gambling in Britain and the USA*, Manchester University Press, Manchester.

Murphy, W.F. (1965) *Wiretapping on Trial: A Case Study in Judicial Process*, Random House, New York, NY.

Murphy, D.E., Kondrashev, S.A., and Bailey, G. (1997) *Battleground Berlin: CIA vs. KGB in the Cold War*, Yale University Press, New Haven, CT.

Murray, E. (1927) *The Post Office*, G.P. Putnam's & Sons, London.

Navasky, V., and Lewin, N. (1973) "Electronic Surveillance", in P. Watters and S. Gillers (eds.) *Investigating the FBI*, Doubleday, Garden City, NY, pp. 297-337.

Negroponte, N. (1995) *Being Digital*, Hodder & Stoughton, London.

Noam, Eli (2014) "United States", in E. Noam (ed.) *Who Owns the World's Media? Media Concentration and Ownership Around the World*, Oxford University Press, New York, NY, pp 500-573.

Nora, S., and Minc, A. (1978) *L' Informatisation de la Societe: Un Réport du Président de la France*, La Documentation Francaise, Paris.

Norton-Taylor, R. (1990) *In Defence of the Realm?*, Civil Liberties Trust, London.

O'Doud, D. (1999) *Consultation Paper on the Interception of Communications in the United Kingdom: HMIC Observations on the Proposed Legislation*, Her Majesty's Inspectorate of Constabulary, London, August (copy on file with author).

O'Harrow, R. (2005) *No Place To Hide*, Free Press, New York, NY.

Olmsted, K.S. (1996) *Challenging the Secret Government: The Post-Watergate Investigations of the CIA and FBI*, The University of North Carolina Press, Chapel Hill, NC, 2nd edition.

Overstreet, H., and Overstreet, B. (1969) *The FBI in Our Open Society*, W.W. Norton & Company, New York, NY.

Parish, A.F.W. (1999) *Interception of Communications in the United Kingdom: A Consultation Paper*, letter addressed to the Home Office Organised and International Crime Directorate, Federation of the Electronics Industry, London, 13 August (copy on file with author).

Parker, E.B. (1973) "Technological Change and the Mass Media", in I. de S. Pool and W. Schramm (eds.) *Handbook of Communication*, Rand McNally, Chicago, IL, pp. 619-645.

Parkhill, D. (1979) "The Necessary Structure", in D. Godfrey and D. Parkhill (eds.) *Gutenberg Two*, Press Porcépic, Toronto.

Parr, M. (1999) *SAQ Response to Interception of Communications*, electronic message addressed to the Home Office Interception of Communications Legislation Team, SAQ Internet, London, 13 August (copy on file with author).

Pearson, T. (1998) *A Response from Internet Service Providers Association and ACPO/ISP/ Government Forum*, letter addressed to Yaman Akdeniz, Cyber Rights & Cyber Liberties, UK, Internet Service Providers Association, London, 02 December, <http://www.cyber-rights.org/privacy/response.htm>.

Perez-Diaz, V.M. (1978) *State, Bureaucracy and Civil Society: A Critical Discussion of the Political Theory of Karl Marx*, Macmillan, Basingstoke.

Perry, C.R. (1977) "The British Experience 1876-1912: The Impact of the Telephone During the Years of Delay", in I. de S. Pool (ed.) "The Impact of the Telephone", MIT Press, Cambridge, MA, pp. 69-96.

Perry, C.R. (1992) *The Victorian Post Office: The Growth of a Bureaucracy*, The Boydell Press/The Royal Historical Society, Woodbridge, Sufolk.

Perry, R. (1999) *Response to Questionnaire for Providers of Communications Services*, London Internet Exchange, Peterborough, 13 August (copy on file with author).

Perry, R. (2000a) *ICF Participating Member Organisations and Personnel*, Internet Crime Forum, Winchester, October, <http://www.internetcrimeforum.org.uk/members.html>.

Perry, R. (2000b) *ICF Terms of Reference*, Internet Crime Forum, Winchester, <http://www.internetcrimeforum.org.uk/homepage.html>.

Pertrow, S. (1994) *Policing Morals: The Metropolitan Police and the Home Office, 1870-1914*, Clarendon Press, Oxford.

Pfaltzgraff, R.L. (1984) "National Security Decision Making: Policy Implications", in R.L. Pfaltzgraff and U. Ra'anan (eds.) *National Security Policy: The Decision Making Process*, Archon Books, Guilford, CT, pp. 291-304.

Phalen, W.J. (2015) How the Telegraph Changed the World, McFarland & Co., Jefferson, North Carolina.

Philips, D., and Storch, R.D. (1994) "Whigs and Coppers: The Grey Ministry's National Police Scheme, 1832", *Historical Research*, No 67(162), p. 77-80.

Pierce, J.R. (1977) "The Telephone and Society in the Past 100 Years", in I. de S. Pool (ed.) *The Impact of the Telephone*, MIT Press, Cambridge, MA, pp. 159-195.

Plate, T., and Darvi, A. (1981) *Secret Police: The Inside Story of a Network of Terror*, Doubleday & Company Inc., Garden City, NY.

Policy Engagement Network (2009) *Briefing on the Interception Modernisation Programme*, The Information Systems and Innovation Group, London School of Economics, Houghton Street, London.

Pollock, D.A. (1973) *Methods of Electronic Audio Surveillance*, Thomas Publishers, Springfield, IL.

Porter, B. (1987) *The Origins of the Vigilant State: The London Metropolitan Police Special Branch Before the First World War*, Weidenfeld & Nicolson, London.

Porter, H. (2009) "Jacqui Smith's Tactical Withdrawal", *The Guardian*, 17 April.

Post Office Engineering Union (1980) *Tapping the Telephone*, Post Office Engineering Union, London, July.

Poster, M. (1990) *The Mode of Information: Poststructuralism in a Social Context*, Polity Press, Cambridge.

Poulsen, K. (1999) "Wiretapping Abuses Alarm EFF, EPIC", *ZDNet News*, Sa Francisco, CA, 21 October, <http://www.zdnet.com/zdnn/stories/news/0,4586,23781,00.html?chkpt=hpqsnewst>.

Pugh, M. (1978) *Electoral Reform in War and Peace, 1906-1918*, Routledge & Kegan Paul, London.

Pye, L.W. (1963) "Communications and Political Articulation", in L.W. Pye (ed.) *Communications and Political Development*, Princeton University Press, Princeton, NJ, pp. 58-63.

Raine, G.E. (1920) *The Nationalisation Peril*, Thornton Butterworth, London.

Rayner, G., and Alleyne, R. (2008) "Council Spy Cases 1,000 a Month", *The Daily Telegraph*, April 14.

Reagan, R. (1989) Speaking My Mind: Selected Speeches, Simon & Schuster, New York, NY.

Richelson, J.T. (1986) *Sword and Shield: The Soviet Intelligence and Security Apparatus*, Ballinger Publishers, Cambridge, MA.

Roberts, N. (1996) *Mobile Phone Crime: How Crime Related to Mobile Phones Can Be Recognised and Investigated*, Federation of Communication Services, London, December.

Robertson, J.H. (1947) *The Story of the Telephone: A History of the Telecommunications Industry of Britain*, Sir Isaac Pitman & Sons, London.

Robertson, K.G. (1982) *Public Secrets: A Study in the Development of Government Secrecy*, Macmillan, Basingstoke.

Rogers, A. (1997) *Secrecy and Power in the British State: A History of the Official Secrets Act*, Pluto Press, London 1997.

Rollins, N. (1992) "The Reichstag Method of Governing? The Attlee Governments and Permanent Economic Controls", in H. Mercer, et al. (eds.) *Labour Governments and Private Industry: The Experience of 1945-1951*, Edinburgh University Press, Edinburgh, pp. 15-36.

Rolph, C.H. (1973) "The British Analogy", in P. Watters and S. Gillers (eds.) *Investigating the FBI*, Doubleday Company, Garden City, NY, pp. 387-411.

Romerstein, H., and Levchenko, S. (1989) *The KGB Against the Main Enemy: How the Soviet Intelligence Service Operates Against the United States*, Lexington Books, Lexington, KY.

Rozenblit, M. (2000) *Security for Telecommunications Network Management*, IEEE Communications Society, New York, NY.

Rubinstein, J. (1973) *City Police*, Farrar, Straus & Ciroux, New York, NY.

Rule, J.B. (1973) *Private Lives and Public Surveillance*, Allen Lane, London.

Russell-Neuman, W. (1991) *The Future of the Mass Audience*, Cambridge University Press, Cambridge.

Russett, B. (1990) *Controlling the Sword: The Democratic Governance of National Security*, Harvard University Press, Cambridge, MA.

Sampson, A. (1973) *The Sovereign State: The Secret History of ITT*, Hodder & Stoughton, London.

Savvas, A. (2008) "Councils Use RIPA Law 10,000 Times to Investigate Citizens for Minor Offences", *ComputerWeekly*, 23 July.

Schramm, W. (1964) *Mass Media and National Development: The Role of Information in the Developing Countries*, Stanford University Press, Stanford, CA.

Schrecker, E. (1986) *No Ivory Tower: McCarthyism and the Universities*, Oxford University Press, New York City, NY.

Sehgal, C. (2000) Electronic message addressed to Joseph Fitsanakis, American Civil Liberties Union, New York, NY, 20 June (copy on file with author).

Seipp, D. (1977) *The Right to Privacy in American History*, Program in Information Resources Policy, Harvard University, Cambridge, MA, June.

Shaiko, R.G. (1998) "Reverse Lobbying: Interest Group Mobilisation from the White House and the Hill", in A.J. Cigler and B.A. Loomis (eds.) *Interest Group Politics*, Congressional Quarterly Press, Washington DC, 5th edition, pp. 255-282.

Shannon, L.W. (1958) "Is Level of Development Related to Capacity for Self-Government?", *American Journal of Economics and Sociology*, 17(4), July, pp. 367-381.

Shannon, L.W. (1959) "Socio-economic Development and Political Status", *Social Problems*, No 6 (4), January, pp. 27-43.

Sims, C. (1999) *Consultation Paper on the Interception of Communications in the United Kingdom*, letter addressed to the Home Office Organised and International Crime Directorate, West Midlands Police, Birmingham, 10 August (copy on file with author).

Smith, P.T. (1985) *Policing Victorian London: Political Policing, Public Order and the London Metropolitan Police*, Greenwood Press, Westport, CT.

Smith, I. (1999) *Interception of Communications in the United Kingdom: A Consultation Paper*, letter addressed to the Home Office Organised and International Crime Directorate, United Kingdom Education and Research Networking Association, Chilton, 12 August (copy on file with author).

Smith Group (2000) *Technical and Cost Issues Associated With Interception of Communications at Certain Communication Service Providers*, Smith Group Ltd., London, 19 April.

Smythe, D.W. (1954) "Some Observations on Communications Theory", *Audio-Visual Communications Review*, 2(4), pp. 24-37.

Snape, T. (2001) "Concerns of Smaller ISPs", presentation given at the *12th Internet Service Providers Association Legal Forum*, London, 05 July.

Spengler, J.J. (1963) "Bureaucracy and Economic Development", in J. LaPalombara (ed.) *Bureaucracy and Political Development*, Princeton University Press, Princeton, NJ, pp. 199-232.

Spiller, P.T., and Vogelsang, I. (1996) "The United Kingdom: A Pacesetter in Regulatory Incentives", in B. Levy and P.T. Spiller (eds.) *Regulations, Institutions and Commitment: Comparative Studies of Telecommunications*, Cambridge University Press, New York, NY, pp. 79-120.

Spindel, B.B. (1968) *The Ominous Ear*, Award House, New York, NY.

Spitzer, S. (1985) "The Rationalisation of Crime Control in Capitalist Society", in S. Cohen and A. Scull (eds.) *Social Control and the State: Historical and Comparative Essays*, Basil Blackwell, Oxford, pp. 312-333.

Srinivasan, K. (2000) "White House Proposes Wiretap Law", *Associated Press*, Washington, DC, 17 July, <http://www.newsday.com/ap/topnews/ap957.htm>.

Stein, J.L. (1996) *Ideology and the Telephone: The Social Reception of a Technology, London 1876-1920*, PhD thesis, University College London.

Steinhardt, B. (1995) *What is the FBI Up To?*, American Civil Liberties Union, Washington, DC, December, <http://www.aclu.org/issues/security/oped.html>.

Sterling, C.H., et al. (2007) *Military Communications: From Ancient Times to the 21st Century*, ABC-CLIO, Santa Barbara, CA.

Stewart, C. (1999) *Home Office Consultation on the Interception of Communications in the United Kingdom: BT View*, British Telecom, Plc., London, 18 August (copy on file with author).

Stoddart, J. (1999) *IOCA Consultation Paper*, letter addressed to the Home Office Organised and International Security Liaison Unit, Lincolnshire Police Headquarters, Lincoln, 02 August (copy on file with author).

Stone, A. (1989) *Wrong Number: The Breakup of AT&T*, Basic Books, New York, NY.

Strum, P. (1999) *To Be Let Alone: Privacy in the United States*, American Civil Liberties Union, <http://www.aclu.org/aclu-e/salon/course5_strum1.html>.

STTCPUSHRCC (2000) *Hearing on Foreign Government Ownership of American Telecommunications Companies*, HR Hrg. 106-153, Subcommittee on Telecommunications, Trade and Consumer Protection of the United States House of Representatives Committee on Commerce, 106th Congress, 2nd Session, United States Government Printing Office, Washington, DC, 07 September.

Sutter, G. (2001) "A Tale of Two Interception Regimes: RIP v CALEA, a Comparison", paper presented at the *16th BILETA Annual Conference*, University of Edinburgh, 09-10 April.

Taylor, D. (1997) *The New Police in Nineteenth-Century England: Crime, Conflict and Control*, Manchester University Press, Manchester.

Temin, P., and Galambos, L. (1988) *The Fall of the Bell System: A Study in Prices and Politics*, Cambridge University Press, Cambridge, 2nd edition.

Teske, P.E. (1990) *After Divestiture: The Political Economy of State Telecommunications Regulation*, State University of New York Press, Albany, NY.

The Economist (1995) *The Death of Distance Survey*, London, September.

Theoharis, A.G. (1971) "Misleading the Presidents: Thirty Years of Wiretapping", *The Nation*, 24 (212), 14 June, pp. 744-750.

Theoharis, A.G. (1978) *Spying on Americans: Political Surveillance from Hoover to the Huston Plan*, Temple University Press, Philadelphia, PA.

Theoharis, A., and Cox, J. (1988) *The Boss: J. Edgar Hoover and the Great American Inquisition*, Temple University Press, Philadelphia, PA.

Thompson, E.P. (1968) *The Making of the English Working Class*, Penguin Books, Harmondsworth.

Tivey, L.J. (1966) *Nationalisation in British Industry*, Johathan Cape, London.

Todd, P.A. (1999) *Consultation Paper on the Interception of Communications in the United Kingdom*, letter addressed to the Home Office Organised and International Crime Directorate, Gloucestershire Constabulary, Cheltenham, 27 July (copy on file with author).

Toffler, A. (1980) *The Third Wave*, Pan Books, London.

Trope, K.L. (2014) "US Government Eavesdropping on Electronic Communications: Where Are We Going?", *The SciTech Lawyer*, 10(2), Winter, pp. 2-8.

Tunstall, J. (1986) *Communications Deregulation: The Unleashing of America's Communications Industry*, Basil Blackwell, Oxford.

Ungar, S.J. (1975) *Federal Bureau of Investigation*, Little, Brown & Company, Boston, MA.

United States Court of Appeals (2000) *USTA v. FCC and USA Respondents On Petitions for Review of an Order of the FCC*, District of Columbia Circuit, Washington, DC, 15 August.

United States House of Representatives Committee on Government Operations (1987) *Report on the Computer Security Act of 1987*, House Report 100-153, Part 2, 100th Congress, First Session, Washington DC, 25, 26 February, 17 March.

USCCGO (1974) *Telephone Monitoring Practices by Federal Agencies*, HR Hrg., United States Congressional Committee on Governmental Operations, 82nd Congress, Government Printing Office, Washington, DC, July.

USSSCSGORIA (1976a) *Intelligence Activities and the Rights of Americans: Final Report, Book II*, Report 94-755, United States Senate Select Committee to Study Governmental Operations with Respect to Intelligence Activities, 94th Congress, 2nd Session, Washington DC, 23 April.

USSSCSGORIA (1976b) *Supplementary Staff Reports on Intelligence Activities and the Rights of Americans: Final Report, Book III*, Report 94-755, United States Senate Select Committee to Study Governmental Operations with Respect to Intelligence Activities, 94th Congress, 2nd Session, Washington DC, 23 April.

USTIA and USEIA (2000) *Lawfully Authorised Electronic Surveillance: Revised Version to Meet the Requirements Defined in FCC 99-230, CC Docket No. 97-213*, J-STD-025A, TR-45, United States Telecommunications Industry Association and United States Electronics Industry Association, Chicago, IL, May.

USTIA and USEIA (1997) *Lawfully Authorised Electronic Surveillance: Interim Standard Version*, J-STD-025, TR-45, United States Telecommunications Industry Association and United States Electronics Industry Association, Chicago, IL, November.

Veigas, H. (1999) *Consultation Paper on the Interception of Communications in the United Kingdom*, letter addressed to the Home Office Organised and International Crime Directorate, Derbyshire Constabulary, Ripley, 27 July (copy on file with author).

Veigas, H.S. (2000) *Regulation of Investigatory Powers Act*, letter addressed to Joseph Fitsanakis, Derbyshire Constabulary, Ripley, 10 November (document on file with author).

Veljanovski, C. (1991) *The Future of Industry Regulation in the UK: A Report of an Independent Inquiry*, Lexecon, London.

Verrier, A. (1983) *Through the Looking Glass: British Foreign Policy in an Age of Illusions*, Jonathan Cape, London.

Verton, D. (2001) "NSA Warns It Can't Keep Up With Rapid Changes in IT", *InfoWorld*, San Mateo, CA, 19 February, <http://iwsun4.infoworld.com/articles/hn/xml/01/02/19/010219hnnsa.xml>

Vesely, R. (1997) "House Panel Questions FBI's Stance on Tap Law", *Wired News*, San Francisco, CA, 23 October, <http://www.wired.com/news/politics/0,1283,7932, 00.html>

Virgo, P. (1999) *IMIS Draft Response to the Home Office Consultation Paper on the Interception of Communications in the United Kingdom*, letter addressed to the Home Office Interception Legislation Team, IMIS, London, 13 August (copy on file with author).

Walker, J.T. (1997) "Re-Blueing the Police: Technological Changes and Law Enforcement Practices", in M.L. Dantzker (ed.) *Contemporary Policing: Personnel, Issues and Trends*, Butterworth-Heinemann, Boston, MA, pp. 257-276.

Ward, M. (2001) "Cybercops Arrest Online Liberty", *BBC News*, London, 18 April, <http://news.bbc.co.uk/low/english/sci/tech/newsid_1283000/1283127.stm>

Watson, D. (2000) *Regulation of Investigatory Powers Bill*, letter addressed to Joseph Fitsanakis, Lothian and Borders Police, Edinburgh, 14 August (document on file with author).

Weatherall, S. (1999) *Interception of Communications in the United Kingdom: Response by Unipalm Group Plc., Trading as UUNet*, UUNet, Cambridge, August (copy on file with author).

Weber, M. (1968) "The Essentials of Bureaucratic Organisation: An Ideal Type Construction", in R.K. Merton et al. (eds.) *Reader in Bureaucracy*, The Free Press, New York City, NY, pp. 18-27.

Weinberger, B. (1995) *The Best Police in the World: An Oral History of English Policing from the 1930s to the 1960s*, Scholar Press, Aldershot.

West, N. (1983) *A Matter of Trust: MI5 1945-1972*, Coronet, London.

Westin, A.F. (1962) "Bookies and 'Bugs' in California: Judicial Control of Police Practices", in A.F. Westin (ed.) *The Uses of Power: Seven Cases in American Politics*, Harcourt, Brace & World, New York, NY, pp. 117-172.

Whitaker, B. (1964) *The Police*, Eye & Spottiswoode, London.

White, T.H. (1975) *Breach of Faith: The Fall of Richard Nixon*, Atheneum Publishers, New York, NY.

White, R. (1999) *Consultation Paper on the Interception of Communications In the United Kingdom*, letter addressed to the Home Office Organised and International Crime Directorate, Powys Police Constabulary, Carmarthen, 23 July (copy on file with author).

Wigham, E. (1976) *Strikes and the Government, 1893-1974*, Macmillan Press, London.

Willing, R. (2000) "New Devices Prompt More Wiretapping", *USA Today*, 03 May.

Wills, G. (1999) *A Necessary Evil: A History of American Distrust of Government*, Simon & Schuster, New York City, NY.

Wingfield, J. (1984) *Bugging: A Complete Survey of Electronic Surveillance Today*, Robert Hale, London.

Wise, D. (1976) *The American Police State: The Government Against the People*, Random House, New York, NY.

Wolmer, V. (1932) *Post Office Reform: Its Importance and Practicability*, Ivor Nicholson, London.

Woods, B. (1993) *Communication Technology and the Development of People*, Routledge, London, 1993.

Yarbrough, D. (1999) *Declaration Before the Federal Communications Commission in the Matter of CALEA*, CC Docket No 97-213, Federal Bureau of Investigation, Washington, DC, 27 January.

Yarbrough, D. (2000) *Communications Assistance for Law Enforcement Act*, Federal Bureau of Investigation, CALEA Implementation Section presentation, Chantilly, VA, October 2000 (copy on file with author).

Printed in the United States
by Baker & Taylor Publisher Services